# High-speed Machining

# High-speed Machining

### BERT P. ERDEL

Society of
Manufacturing
Engineers

Machining Technology
Association of SME

**Dearborn, Michigan**

Library of Congress Catalog Card Number: 2003103270
International Standard Book Number: 0-87263-649-6

Additional copies may be obtained by contacting:
Society of Manufacturing Engineers
Customer Service
One SME Drive, P.O. Box 930
Dearborn, Michigan 48121
1-800-733-4763
www.sme.org

SME staff who participated in producing this book:
Phil Mitchell, Senior Editor
Rosemary Csizmadia, Production Supervisor
Frances Kania, Production Assistant
Kathye Quirk, Graphic Designer/Cover Design
Jon Newberg, Production Editor

Cover photo courtesy Mazak Corp.

Printed in the United States of America

## ABOUT SME

The Society of Manufacturing Engineers (SME) is the world's leading professional society supporting lifelong manufacturing education. Through our member programs, publications, expositions and professional development resources, SME promotes an increased awareness of manufacturing engineering and helps keep manufacturing professionals up to date on leading trends and technologies.

Headquartered in Dearborn, Michigan, SME influences more than half a million manufacturing engineers, executives, and practitioners annually. The Society has members in 70 countries and is supported by a network of hundreds of chapters worldwide.

## ABOUT MTA/SME

The Machining Technology Association of SME (MTA/SME) provides a forum for exploring the practical application of innovations in material-removal processes, such as turning, milling, and grinding, and the machines and tooling used in these processes. MTA/SME members represent a cross-section of disciplines, including manufacturing, tool, process, mechanical, and industrial engineers, production and plant managers, and researchers and educators. They are interested in the latest developments in CNC machine tools, cutting-tool materials, tool geometry, workholding devices, and cutting fluids. Typical industries served include automotive, aerospace, defense, appliance, and other general metalworking manufacturing sectors.

*To My Beloved Children*
*Christoph, Gillian, and Wolfgang*

# Acknowledgments

This book would not have been possible without the extraordinary support of my staff and team who have supported me over the years on our exceptional path of success. To all of you I am deeply indebted.

My special thanks go to Dr. Michael Kaiser for his technical assistance and expertise; to Susan Ephraim for her patience and diligence in formatting; and to Dave Itterly for the outstanding artwork. I would also like to thank Jerry Scherer at GE Fanuc Automation for reviewing certain chapters of this book.

I express my deep gratitude to the many people who have inspired me over the years. They have been instrumental in writing this book in so many ways.

# Table of Contents

# Preface

Manufacturers are intertwined in a global network of seemingly endless streams of information and product varieties, and as life picks up pace, producers and consumers turn to scientists and engineers for technology-based innovation. The question is whether continued innovation can be sustained enough to maintain manufacturing progress, and whether manufacturing companies have the acumen to adopt new strategies, systems, and processes. The growing demands of customers are transforming the way products are manufactured.

This book is written to show how high-speed machining technology and innovation can be used for the advantage of all industrial manufacturing. It is possible to economically produce exceptional quality products with high productivity. However, it can be done only if companies take a holistic approach. They must assemble the collective knowledge of all facets of the enterprise and apply it to current processes and technologies for their continuous improvement.

Manufacturing entities can no longer afford to just look at machines and tools for producing products. Manufacturing paradigms will continue to gravitate toward agility, rapid product output, shorter product development cycles, and the ability to adjust to erratic market changes, as well as to a company's own innovative abilities and corporate knowledge.

This book will show how to secure high performance on the production floor and explain its benefits. Moreover, it will show that manufacturing has not yet reached the point of diminishing returns. If companies embrace technological innovation, the challenges ahead are exciting opportunities in the world of manufacturing.

# Introduction

**1**

There is a belief that competitiveness, especially on a global scale, helps cut waste, maintain high quality standards, and generally fosters the "lean" manufacturing paradigm. Improving operations throughout a manufacturing operation is a task of perpetual consequence. What might be good enough today, most probably will not be tomorrow. New challenges become opportunities for improvement.

The series of activities that take place as raw materials are transferred into finished product and sold to the end user is called the *value chain. Outsourcing,* the strategy that lets original equipment manufacturers (OEMs) concentrate on what they do best (core competency) while shedding off what others can do better, has changed the manufacturing rules. Collaboration with suppliers and partnership arrangements has led to the extended enterprise.

Continued innovation can sustain the momentum of manufacturing progress and manufacturing companies have the capabilities and business acumen to adopt new strategies, systems, and processes. Speed and accessibility of information, global reach, and closeness have dramatically altered our way of life. The impact is great as environmental concerns and health issues have begun to transform the way things are manufactured. The growing demands of customers are rewriting the current way of thinking. In the past two decades, the affect of manufacturing on the environment has become a growing concern. The upshot: products must be made with the least possible harm to the environment. In some cases, a reduction of energy usage, substitutions for traditional material, different designs, applying new machining technology, and promoting use of recyclable and biodegradable materials can have an enormous impact on securing a protected, healthy environment. Some governmental laws and restrictions are already in place.

Any responsible manufacturer has to deal with environment issues to be successful. Behind the challenges lie untapped opportunities for new technology, product lines, and markets, provided environmental considerations become an integrated part of the enterprise.

Cost, quality, time, and continuous improvement have become the basis for the pursuit of best practices—the fundamental yardstick of today's manufacturing enterprise. The bottom line for corporate success is customer satisfaction. Customer purchases are at the center of it all. The performance of the acquired products has everything to do with the way they are made. Manufacturing has to embrace process much more so than in the past. Processes are a wide-open playing field where the adoption of advanced and innovative technology can open up and create new markets. After all, it is the process that makes the product.

U.S. manufacturers are adopting agile manufacturing principles to achieve product variety in style and volume within the shortest possible changeover times to minimize the response time to demanding customers.

For the manufacturing floor, agility means to be ultra-flexible in tooling and automation. Cycle-time reduction, fast throughput, high machine uptime, minimum machining passes, predictable machining results, and good part designs are at the core. In addition, machining processes have to be adaptable to transfer lines and computer numerical control (CNC) cells or a combination thereof, and if possible, reusable for new projects to assure a short and economical pipeline to the next generation of products.

## ADVANCED MACHINING PROCESSES

High-speed machining, also called high-velocity machining, is often specified for new investments in manufacturing, especially when machining nonferrous metals. This makes sense, given the progress in machine spindle technology and the velocity at which the machine axes can travel. In conjunction with advanced cutting materials, particularly polycrystalline diamonds, high-speed machining substantially reduces machining and non-production time.

The situation is different when machining ferrous metal. Cast iron, carbon and alloy steels, as well as super alloys have their specific characteristics. The following characteristics make it difficult to apply high cutting speeds.

- abrasive pearlite structures of nodular and ductile irons;
- hard particles in powdered metals;
- austenitic (stainless) steels with poor thermal conductivity; and
- nickel, chromium, and titanium-based alloys are of extreme hot-hardness.

Since it is either technologically not feasible and/or impractical because of an unfavorable price/performance ratio to apply cutting speeds close to those applicable for nonferrous metals, other options have to be used. Besides the cutting speed, attention has to be paid to minimizing machining time and optimizing the entire machining process. This is what is called high-performance machining. High-performance machining does not necessarily require the acquisition of new machinery; the efforts go toward improving existing processes. The most promising and intriguing part of the process is the cutting tool system, because it can be easily adapted to specific applications. When the application drives the process, the cutting tool system mostly determines the level of machining and manufacturing performance. The time it takes to get a product out the door depends on how fast the parts are machined and the security and maturity of the process. Cutting tools have to be highly productive and low cost. New tooling systems use breakthrough technology, which produces high-quality parts machined with speed and zero defects.

# A NEW ERA

There is a definite urgency to retool for a new era of fierce competition. More production flexibility and agility is needed to accommodate product mixes, varying production volume, and broader part families.

Processes that sustain a high degree of productivity will dominate if they signify cost-effectiveness and guarantee enhanced part quality. The key is the process—the entire machining envelope—not individual cutting tools or a particular machine or isolated coolant issues. Today, a greater emphasis is put on technologically advanced processes, which include high-speed/high-velocity machining. Exploiting every technical potential, scrutinizing every technological angle, and transforming them into productive use will produce heretofore unheard of manufacturing results. High-performance processes not only elevate day-to-day manufacturing to different heights, but they open up new competencies and the chance for technology leadership. The following chapters will describe the pathway to such technologically advanced processes and illustrate their superior benefits.

# The Determinants of High-speed/ High-performance Machining

**2**

The needs for energy and weight savings complement the manufacturer's directions for lighter and smaller product designs. However, there is also the demand from the end user and consumer for more built-in features, notably in the areas of safety, communication, and comfort. Smaller and lighter seems to be diametrically opposed to more content. However, these criteria can balance each other through smart design and manufacturing.

Today's automobile is a prime example. Considering the increase in performance, reliability, and available comfort and convenience options, the car's weight reduction is remarkable. New cars typically have smaller, lighter engines, and transmissions. Yet, output and reliability are higher. Anti-lock brake and anti-slip systems are add-ons, as are front and side airbags. Smart designs and improvements in technology combined with advanced manufacturing processes make these innovations happen. Other obvious examples of performance increases and smarter products in line with smaller, lighter designs are household appliances and computers. Product users do not want to sacrifice comfort, safety, or health because of product size and weight.

It appears that there is a continuing need for more complexity in services and products offered. Nevertheless, in spite of a more complex world, if functions of processes and products are too complicated, they become unacceptable, especially if potential alternatives are simpler. Speed is an essential enabler of success at any company. It has to permeate the entire organization from product development to production, to delivery to the customer, and every step in between.

## WEIGHT, MATERIALS, MACHINE TOOLS

The potpourri of lightweight products is increasing. The conversion rate from heavy to lighter weight will accelerate exponentially over the coming years and it will change the way products are manufactured and machined. The selection of material is dependent on:

- machinability,
- formability,
- durability,
- weight,
- physical characteristics,
- chemical composition,
- safety considerations,
- cost (purchasing, usage, and disposal),
- environmental issues (recycling and disposal), and
- availability.

The automotive industry has always spearheaded certain technology and has been a trendsetter of things to come. The major drivers of new car technology are:

- stringent exhaust emissions,
- low mass,
- fuel economy,
- ease of handling,
- material strengths,
- comfort, and
- safety.

Spurred by the need for lightweight cars to improve fuel economy, automakers are feverishly working on material that can cut the average car weight to 2,000 lb (907 kg), down from 3,300–3,500 lb (1,497–1,588 kg) today. The current mix of materials used for automobiles will change rapidly, as shown in Figure 2-1.

Composites, aluminum, and magnesium will see the most dramatic increases, the latter two mostly for powertrain design. Composites and plastics will challenge steel components, although steel mills are scrambling to keep their share by developing ultra-lightweight steel that can accommodate the expected weight reduc-

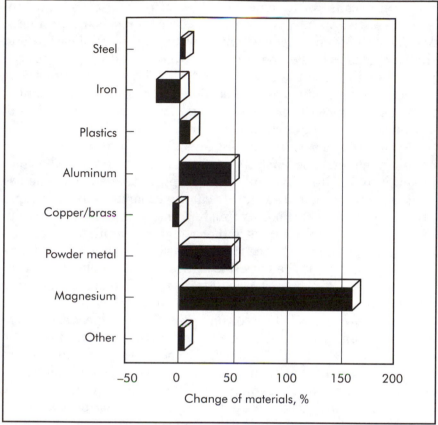

Figure 2-1. Materials used in typical passenger cars.

tions. The distinct advantage of the steel is its recyclability. A *composite* is typically a composition of fibers and binding material that can be made with high desired strength and durability for lightweight structures. As soon as composites manufacture becomes more economical, their share will go up in almost all product manufacturing.

The key to putting a product on a diet is to redesign it and specify substitute materials. Both approaches have their limitations. Design is determined by functionality. Economics and physical characteristics drive the search for other materials. For example, there is aluminum's success as a lightweight component material because of its amenable cost of purchase and machinability, as well as physical characteristics for specific products. Materials have to be paired with the optimum application and that includes cost. The need for lightweight and weight-reduced products will spark the development of new production methods to accommodate advanced materials. There is simply no turning back the tide towards lightweight products. Major contributors of advanced material growth are described in Table 2-1.

Getting the ideal material for specific applications is still an ongoing challenge because the process requires mixing the components that will satisfy a particular application's need. The use of lighter material, more accurate castings, parts made from rapid prototyping, and workpieces finished in a single machining operation all but eliminate pre- and/or rough machining. Heavy cuts are no longer needed, hence the decrease of cutting forces necessary to cope with those cuts. Instead, machine tools offer other much needed features such as agility, open architecture, high axes acceleration rates, high spindle speeds, and zero cut-to-cut time. They are now lightweight and conveniently movable within dedicated machining areas. Machine tools also offer a more favorable price/performance ratio. Lightweight machine tools do not require concrete slabs, but can be mounted on any typical manufacturing floor. They are built for dynamic stiffness and economy.

Optimized castings of unfinished parts will contribute as much to lighter parts as will fabricating finished parts through stamping. In all, the entire manufacturing envelope will shrink. Product-life cycles will become shorter and shorter, along with the times allowed for product development. Manufacturers have to respond

Table 2-1. Material uses by industry

| Material | Main Usage |
| --- | --- |
| Plastics | Household goods, automotive |
| Composites | Automotive, aerospace |
| Aluminum | All industrial manufacture |
| Magnesium | Automotive, aerospace |
| Powder metal | Automotive |

to market changes swiftly and vigorously. This translates to rapid innovation, speedy marketing, and fast production.

Although demanding more gadgets and options, which add to product complexity, end-users want to handle and operate products with ease and simplicity. Interaction of man and machine, whether on the production floor or in the post-production stage and in the market, increase the desire to equip machinery with elevated "brainpower."

## SIMPLE PROCESSES AND SYSTEMS

The manufacturing process, handling and operation of products, communication, as well as product and technology accessibility, will need to be simple. Even with efforts underway to standardize within product mixes, the variety of products in industrial manufacture is overwhelming. Consequently, manufacturing processes vary from plant to plant. For example, lights-out production, where second and third shifts produce automatically without human interface, will depend highly on product variety, product unit, part families, and complexity. This makes a clear case for simpler processes. While complete robotic automation might seem to be simpler, because it takes the human element out, it certainly puts a tremendous burden on monitoring, safety margins, and maintenance procedures.

The much-heralded computer-integrated manufacturing (CIM) of the 1980s has long made room for being more efficient. The results are simpler manufacturing and machining methods. Rigidly installed machine tools, interconnected through a dedicated part transportation system, have been replaced by machining centers that can be arranged and rearranged as desired. The "plug and play" principle and with it "agile" manufacturing, offer much simpler manufacturing processes. They can be easily monitored and accommodate part changes and increase floor efficiency through simplicity. These same systems will play an important role for "virtual" production as well as concurrent manufacturing. For the latter, it will allow easy duplication of original equipment manufacturer's (OEM's) international facilities, because of its functional and operational simplicity.

As to real virtual manufacturing, industrial enterprises will erect manufacturing places on a limited-time basis and keep those plants under power only for the duration of the time frame that certain parts and products are in demand. Those facilities may work to produce an overflow capacity of parts (otherwise made at different plants to accommodate peak demands) or products, whose existence in the market are limited according to specific market indicators. Process simplicity invites manufacturing agility, which must be pursued to stay competitive.

Machining processes primarily consist of machine tools and cutting tools. While tooling traditionally only makes up less than 20% of the total production cost, it conversely directly influences the other 80% of the production cost. The right tooling, not necessarily the machine, often makes the difference in productivity on the production floor. Manufacturing cannot afford to specify machine-tool simplicity without the same stringent requirements on tooling—an area still widely untapped. The technology that provides both precision and simplicity must be made more available. Included in this are:

- one-pass finish machining;
- no post-machining processes;
- using tools for regular production machines;
- tools made by modular design;
- ready-made tools out of stock;
- tools with repeatability accuracy;
- tooling commensurate with finish requirements;
- tools suited for high-speed machining;
- easy and reliable precision tool clamping;
- using cutting material with the best price/performance ratio;
- eliminating noncutting time and tool setup time;
- defined tool management; and
- avoiding excessive heat during cutting and seeking short-curled cutting chips.

Specifying advanced cutting-tool technology will make machining processes simple and more efficient. American manufacturing companies are spending large amounts of money to acquire top-notch machinery. However, most of the tooling in use will have to be scrapped. The fundamental approach to a manufacturing process will no longer be to acquire the latest, futuristic machine tools and fit them with outdated tooling. All links of the manufacturing chain have to be in line with the latest technology if they are to be of measurable strength in simplicity and operation. On the product end, designers will specify and create products for ease of manufacturing (design for manufacturability) and ease-of-operation and handling, and product access for the consumer (end user).

Simplicity in product identity and access will have manufacturers looking more at their own operations. Purchasing decisions will be made as the result of direct sales. OEMs will be closer to their clientele and will be no-nonsense and much simpler. For example, a car manufacturer will have direct access to feedback from the customer, allowing it to produce products individually, based on customer specification. In the future, image building in the automotive industry will be the strategy of choice. The simpler the OEM's setup and relationship with its customers, the faster and more beneficial the outcome.

Of course, easy, reliable communication is a relevant tool for process and product. In fact, it is the enabler for reliable manufacturing and marketing. Internet communications offer the manufacturer, and its customer, a simple, fast and inexpensive interface. Beyond that, it enables suppliers to the OEM to assist and support their equipment and services. For processes, the OEM, machinery, and tooling supplier can interact in their daily activities on the OEM's shop floor. Connected with machines and tooling through video and audio systems, two-way communication allows for application assistance, troubleshooting, diagnostics, process monitoring, and operator training. PC-based, bi-directional communication links the OEM and supplier through commonly available hardware and software.

## FAST MACHINING, RESPONSE TIME, AND THROUGHPUT

Buyers will demand rapid changes in the products available in the market. Consumers' tastes will shift more erratically and often unpredictably. The common denominator for manufacturers to satisfy their customers along these lines is speed. It means developing new products fast, accelerating production rates, and shortening time to market as much as possible.

The old sequential methods of product development that ruled almost the entire 20th century have no more merit. Product complexity, superceding technology, and the need for input from all corporate branches mandate developing products through cross-functional teams by way of concurrent engineering principles, as shown in Figure 2-2.

When Chrysler developed the Viper® sports car, it took them only 2-½ years from concept to product, cutting their typical product development time in half. Considering the limited number of people on the team and, the resulting new products and manufacturing principles, it was quite an achievement. While it set new standards for American car development timetables, Toyota, roughly five years later, announced that it cut the time it took to develop a product to only one year.

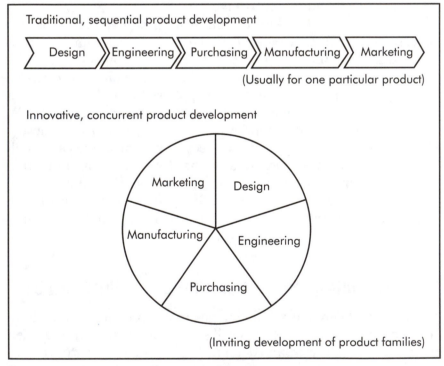

*Figure 2-2. Product development methods.*

Increasingly faster product development is spawning relentless pressure to accelerate production tasks and cut operating costs. High-speed/high-velocity machining reduces cycle time by shortening main machining time. Manufacturing must specify the characteristics necessary to take advantage of this significant technological advancement. If machine tool, tooling interface, and cutting tools are commensurate in technology degree, high-speed/high-velocity machining answers the quest for lower machining time and cycle time.

To complete the cycle, products are brought to market and made available to the end-user, the buyer of the products, in the speediest way possible. Depending on the type of product, this is done through dealers, agencies, shopping outlets, malls, and distributors or direct from the manufacturer. The distribution channels will undoubtedly change. Manufacturers will have to know directly from customers what products they prefer. This will necessitate customers having direct lines of communication to the manufacturer with the ability to order from product menus (Figure 2-3) (Lutz 1998).

Over time, the manufacturer will be able to position itself better for standard equipment by proactively narrowing down the scope of the supply chain. Reading the market accurately is an essential aspect of bringing products to market rapidly.

Market dynamics are such that the only constant seems to be change. Grasping, handling, and executing them throughout the manufacturing enterprise is essential. Information systems, such as computer networks based on manufacturing software modules and built on the premise of pleasing customers, must handle agility and speed. Such systems must provide direct access to manufacturing for quick product turnaround and communication of available capabilities. As the organization receives the order, an assessment has to be made as to the availability of labor, material, and processes. The order is seen in real time and allows for immediate adjustments or changes to production as well as distribution schedules. The summation of the inquiries is condensed to inform the customer of the exact status of the order instantly. Having fast in-house information systems will no longer be enough; their capabilities have to be communicated to the outside world—the customer.

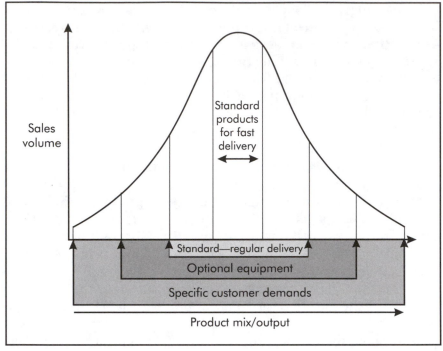

*Figure 2-3. Product mix compared to sales volume.*

One of the premier attributes of successful corporations will be swiftness. Conversely speaking—slowness will be deadly. Corporations' activities must be speedy, so objectives can be met quickly.

## SMART MACHINES, TOOLS, AND PROCESSES

Manufacturing companies will have to develop, more than before, built-in intelligence to get a firm grip on manufacturing predictability, and automated results, as well as flexibility for both process and product, as shown in Figure 2-4. The guesswork must be taken out of the systems.

Parts production will have to be done reliably through integrated systems. Reducing or increasing part production volume and variety has to be performed smoothly. At the core of it is *agility*—the intellectual and technological preparedness for an-

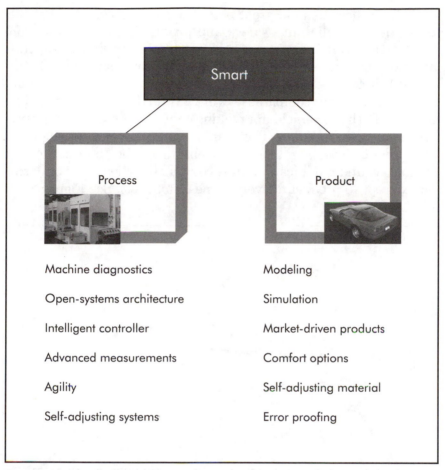

*Figure 2-4. Smart process and product.*

ticipated and unanticipated changes. From the manufacturing point of view, it means to provide intelligent communications within processes. Such communication includes standardized open-systems architecture and the integration of the factory floor in terms of machining operations and data through intelligent controllers. Remote machine and process diagnostics can be sent from the supplier to the end-user via computer interface to monitor and correct machining results. In-process inspection and monitoring through sensor-based control, as well as advanced measurement technologies, are parts of it, too.

Smart processes hold great potential to improve upon productivity, quality, reliability, safety, and production economics. Continued research in industry and academia for more real-time autonomous processes will take the quest for smart manufacturing full circle. Modeling and simulation—two essential techniques for manufacturing companies—will gain further importance (Figure 2-5). Both are vehicles for making confident predictions for process and product design. They can substantially cut development and redesign costs, and reduce or eliminate the time needed for evaluation and modification, thus condensing the time from product concept to product delivery (time to market). See Figure 2-6.

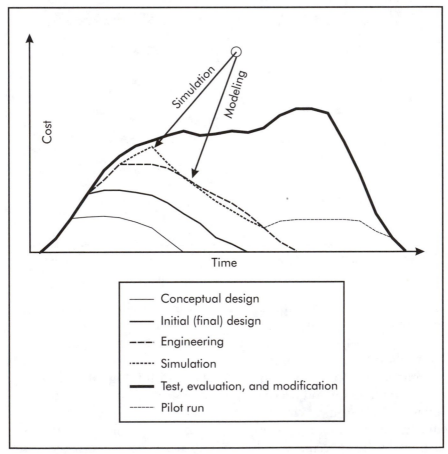

Figure 2-5. Simulation and modeling.

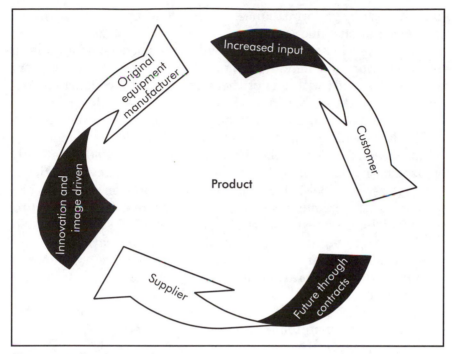

*Figure 2-6. Product cycle.*

Tools for simulation and modeling vary from 3-D computer technology to rapid prototyping to one-time casting. Ultimately, modeling and simulation will be part of an enterprise information system interconnected to all relevant department workstations and databases. This will enable the manufacturing entity to design, correct, optimize, and validate within a virtual framework. It will provide scenarios of understanding and evaluation of best process and best product with certainty.

Modeling and simulation enhance the ability of companies to manufacture market-driven products. A typical example is Chrysler's Viper sports car. Answering the customers' call for a high-powered two-seater, simulation and modeling techniques cut the traditional development time in half and created a market-driven product—timely and with great success.

A manufacturer in the 21st century can, and will, mostly produce products that have their origin in the market and answer customers' demands and needs. Only a few manufacturers will have the courage to introduce a product without market forces and without customer demand.

Clearly, products progressively will feature options for enhancing operation and handling comfort. The automotive industry, undeniably faced with one of the most competitive markets, is a harbinger of things to come, and of options that will spill over to other industries, notably household and computer products.

Products will include proactive controls to detect obstacles or hindrances before they occur and adjust to changes automatically. Noise-abating material will silence consumer products, practically down to zero decibels. Operational buttons for even the most complex goods will be narrowed down to only a few. Glass that automatically blackens when cars are locked can hide what is inside. Material that adjusts itself depending on temperature and mechanical stress in aircraft will make flying safer. Pre-programmed features will set operating functions in motion without human interference. While this all enhances the end-user's comfort and safety, it makes products more complex. Even so, the customer expects it to be maintenance and trouble free. Hence the never-ending strive for zero defects.

A good part of producers' activities will be devoted to error proofing, mostly involving computer-enhanced validations (poka-yoke). With it, the manufacturing floor can identify weak spots and possible problems. Automatic quality checking systems can be installed at the respective assembly-line station to capture a faulty part before it goes further on the assembly line.

To sum it up, in the $21^{st}$ century, processes and products alike will be described as small, light, fast, simple, and smart (Figure 2-7).

# REFERENCE

Lutz, Bob. 1998. *Guts: The Seven Laws of Business that Made Chrysler the Hottest Car Company.* New York: John Wiley and Sons.

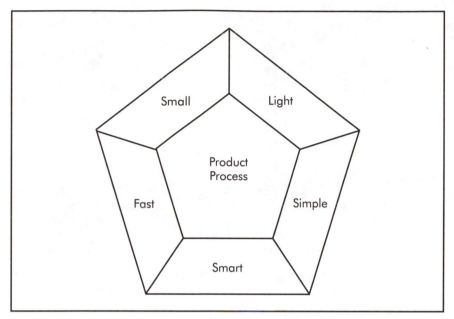

*Figure 2-7. Requirements for process and product.*

# Characteristics of High-speed Machining

**3**

Processes that can sustain high degrees of productivity will dominate if they offer the potential to enhance quality while remaining cost effective. It is here that the complex technology of high-speed/high-velocity machining is extremely promising. By synthesizing a myriad of technologies, high-speed machining reduces cycle times, increases throughput, minimizes non-machining time, shortens main machining time, and yields high precision with a favorable cost/performance ratio.

Today, *high speed*, as defined by the rotational spindle speed, spans a range of 15,000–25,000 rpm for typical machining operations. However, since the tool translates rotational (idle) spindle speed into practical use, the term "high cutting speed" is more appropriate. High cutting speed ranges between 914–1,524 ft/min (279–465 m/min) in nonferrous applications (mostly aluminum parts). However, the process also has been applied to steel and cast iron when cutting at speeds of 305–366 ft/min (93–112 m/min). In any event, the influence of high cutting speeds within the machining envelope is rather dramatic, as shown in Figure 3-1.

## MACHINING PARAMETERS

While chip volume and surface quality increase with higher cutting speeds, cutting-tool life is shortened and cutting forces decrease. A smaller amount of the heat generated at the cutting point is transferred into the workpiece and cutting material than is normal with traditional machining, because much of the heat is removed with the chips. The lower tangential loads on the cutting tool simplify part-fixture design and allow for easy machining of thin-walled workpieces. Better surface finishes are, as a

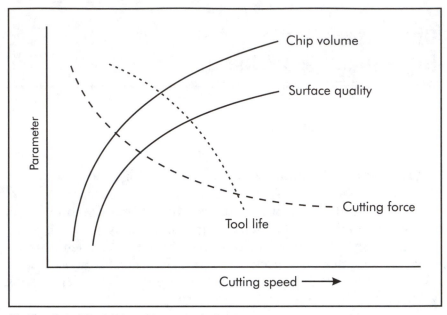

*Figure 3-1. Machining characteristics.*

rule, a welcome bonus, as are the higher chip-load demands pro-vided there are well-defined tool-chip galleys and a reliable trans-port of the chips out of the machining area. Advanced cutting-tool material can prolong tool life, and hard, heat-resistant cutting material can keep reductions of tool life to acceptable levels.

To reap the benefits of high-speed machining, it is important to synthesize the technology that comprises the machine, tooling in-terface, and cutting tool. All of the individual components that make up these three groups must be well designed and engineered to their respective highest levels and, in unison, must represent state-of-the-art design and workmanship. The most relevant cri-teria to consider are shown in Figure 3-2.

High-speed machining allows manufacturing companies to re-spond quickly to sudden market shifts and different product mixes.

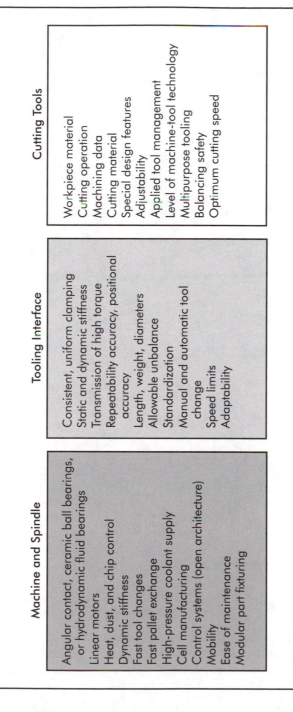

**Machine and Spindle**

Angular contact, ceramic ball bearings, or hydrodynamic fluid bearings
Linear motors
Heat, dust, and chip control
Dynamic stiffness
Fast tool changes
Fast pallet exchange
High-pressure coolant supply
Cell manufacturing
Control systems (open architecture)
Mobility
Ease of maintenance
Modular part fixturing

**Tooling Interface**

Consistent, uniform clamping
Static and dynamic stiffness
Transmission of high torque
Repeatability accuracy, positional accuracy
Length, weight, diameters
Allowable unbalance
Standardization
Manual and automatic tool change
Speed limits
Adaptability

**Cutting Tools**

Workpiece material
Cutting operation
Machining data
Cutting material
Special design features
Adjustability
Applied tool management
Level of machine-tool technology
Multipurpose tooling
Balancing safety
Optimum cutting speed

*Figure 3-2. Machine and spindle, tooling interface, and cutting tools.*

# Machine-tool Technology

# 4

In the past few years, machine tools have undergone dramatic changes. Technology-based innovation has altered their manufacture and application. The results are clearly measurable through substantial productivity progress, manufacturing efficiency and effectiveness, paired with more economical and precise machining.

Machine tools set the stage for high-performance machining through various attributes:

- manufacturing and multi-task machining systems;
- high-speed/high-velocity machining;
- near-dry machining;
- multi-task/multipurpose in one pass; and
- peripheral technology.

## MANUFACTURING AND MULTI-TASK MACHINING SYSTEMS

No question, the machine-tool market has been changing markedly relative to make up, line up, technology, and specification. More and more machining centers are capable of shaping parts to their final form and size. Without sacrificing productivity, lightweight, flexible machines, running at high speed and acceleration rates, combine more machining tasks in one machine. And, they are supported by simple peripheral equipment. Less complicated processes and fewer machines translate into less capital equipment expenditure, making parts machining more economical. To be sure, the automotive industry still acquires transfer lines for some uses. Built-in flex stations and rotary machines are still popular for one- and two-part family runs. However, in the next few years, machining centers will dominate manufacturing, and for good reasons.

A variety of machining centers are available to choose from and their selection is not based on just table size, spindle horsepower, and cut-to-cut time the way it was a decade ago. Table size, spindle output, and the time it takes for tool change still matter. However, the criteria for advanced machine tools are different today. They are built very compact, robust, and powerful. Some are run with extremely high spindle speeds and acceleration, or are equipped for near or completely dry machining. Options for additional tool storage, tool monitoring, and shuttle tables are usually part of the scope of supply as well as single and twin spindles. Vertical and horizontal spindle configurations can be specified. Sophisticated control systems with open architecture, interpolation capabilities, and a host of subprograms are in demand for complex part machining. There are developments in the marketplace and demands by end users that drive more innovative solutions and new technology to convert the single, stand-alone machining center into more of a multi-task machining and manufacturing system, as shown in Figure 4-1.

The machining center has become the last link of a manufacturing chain that has more upstream links and activities. Downstream, there should not be much left except for final inspection since post-machining processes are mostly eliminated. This is due to the pursuit of one-pass machining and multi-tasking within any machine envelope. To make that happen, new developments are taking place upstream that directly influence the specifications and task of the machining center, as shown in Table 4-1.

The downstream developments that affect the characteristics of machining centers are shown in Table 4-2.

The perpetual strive for more productivity and high precision through more economical methods and less energy expenditure has spurred radical new machining concepts. The hexapod machine, shown in Figure 4-2, is a complete departure from the conventional basic machining-center design. Going back to its original design for flight simulation, the hexapod, also known as the Stewart platform, has a parallel-link mechanism.

Every machine command is a nonlinear relationship of six coordinates. Every motion has to be translated into six coordinated

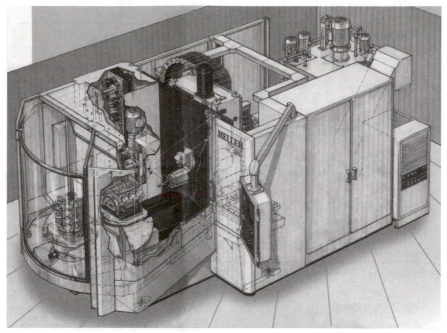

*Figure 4-1. Advanced machining center.* (Courtesy Heller Machine Tools)

## Table 4-1. Upstream developments

| Upstream Developments | Machining Center Task |
| --- | --- |
| (Near) net shape of parts | Lighter cuts |
| Rapid prototyping | Less tryouts |
| Nonferrous part materials | Higher speeds and feeds |
| Modeling and simulation | Secure process |
| Environment and health | Less coolant use |
| Small workpiece design | Ease of handling |
| More complex parts | More tooling, better control system |
| Mobility | Small floor footprints |
| Connectivity | Total integration with other in-company activities |
| Virtual manufacturing | Robust processes |

Table 4-2. Downstream developments

| Downstream Developments | Machining Center Task |
| --- | --- |
| Fast-changing consumer tastes | Flexibility/adaptability |
| Technology preparedness for changes | Advanced technology capabilities |
| Remote machine and process diagnostics for manufacturer to end user | In-process inspection, sensor -based machining/monitoring |
| Less inventory, just-in-time (JIT) delivery | High machine uptime, minimum non-machining time |
| Instant delivery | Robust process |
| Global manufacturing, outsourcing | Open-architecture simplicity |
| Quality awareness | High-precision finishes |

leg lengths, moving in real time. Motor-driven ball screws extend or retract the struts in response to machine-control demand. Coordinated motion of the six struts offers volumetric accuracy, because of the parallel movements.

The hexapod's rigidity stems from the six ball screws that share the load. There is only longitudinal tension and compression stress rather than bending stresses. Also, its lighter weight when compared to conventional machines and low friction make it less prone to settling out. However, the motion of the ball screw generates a lot of heat, affecting the accuracy of the machine due to thermal expansion. This is solved by thermal compensation in the software and the inner cooling of hollow struts, which can lower the generated heat. More intense efforts are needed to get a basic understanding of performance, speed, accuracy, stiffness, monitoring and measuring systems and length compensation, and the transformation of Cartesian coordinates into machine coordinates.

Parallel kinematics link mechanisms have already made their mark in aerospace and the mold and die industry to perform complicated contour machining. Hexapods and three-axes derivatives might revolutionize the way machining is done today. Intense research and development is needed to make these machines the mainstay of tomorrow's manufacturing floors.

*Figure 4-2. Hexapod machine.* (Courtesy Ingersoll International, Inc.)

## Multi-task Machines

Another interesting approach to speedy machining processes and gaining productivity is the realization of bringing nonproduction machining processes into the envelope of production machines. Examples are spark-erosion processes that complement CNC-grinding machines and finish grinding combined with milling machines.

## HIGH-SPEED/HIGH-VELOCITY MACHINING

High speed and high velocity are relative terms. They are relative to the workpiece material, operation, and perspective of the user. Secondly, the rotational speed of the spindle is really only meaningful if complemented by high axes movements. Thirdly, the range of rotational speeds and linear accelerations, where the term high is appropriate, is determined by the technology and design characteristics of the machine. Finally, speed and acceleration have to be tuned to each other intelligently.

## Machine Spindle

There is an array of design elements that contribute to the function and reliability of high-speed spindles (see Figure 4-3). High spindle speeds for production machines start at 10,000 rpm and flatten out at about 20,000 rpm. The design differences can be significant. Accuracy, reliability, temperature, acceleration, balancing, and machine lifetime are factors that determine the design requirements.

## Bearings

For minimal friction as well as high-axial and radial-load capacity, angular, ceramic ball bearings are most suitable. Angular-contact bearings allow for heavier cuts; ceramic bearing material builds light, suffers minimal thermal expansion, and allows for

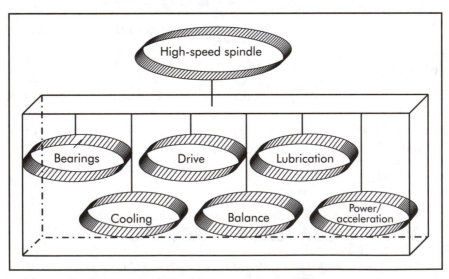

*Figure 4-3. Major modules for high-speed spindles.*

higher spindle pre-load. Today's spindle technology allows machine tools to adjust pre-loads depending on speed and thus centrifugal force (higher speed—higher pre-load). Hydrostatic/hydrodynamic fluid bearings are excellent, but more expensive alternatives. Their main advantage is they inherently provide excellent dampening during high-speed machining and, thus, assist in balanced machining conditions.

## Motor/Drives

Direct connection of the motor to the spindle has prevailed over the use of belts and gears to connect the two. An integral spindle, coupled directly to the motor, accelerates/decelerates faster than a system involving a drive-train mechanism. An integral spindle has a symmetric growth rate along its axis in terms of thermal expansion and is, therefore, more manageable along these lines. The same holds true for vibration control, making spindle balancing much easier.

The technology that has really given a boost to high-speed machining is the linear (motion) drive. Such drives fall into two categories, namely permanent-synchronous-brushless and asynchronous-linear-induction motors. The brushless technology has advantages because of its simple mechanical design and its lesser heat distortion. Linear induction motors are better for longer, continuous magnetic fields and long travels.

Although the cost for linear motor drives is about 50% higher than traditional ball screws, acceptance is escalating simply because of its accuracy and stability during high-speed travels. Linear motor drives produce rapid traverse rates of 246–492 ft/min (75–150 m/min) and accelerations of up to 82 ft/sec$^2$ (25 m/sec$^2$ or 2.5 g). For the lower end of high-speed machining, newly designed ball screws featuring hollow shafts for height and weight, and double ball screw arrangements for accuracy and heat dissipation, foregoing distortion, can still be a solution. But this is only possible in conjunction with advanced control systems capable of processing new algorithms and multiple, dedicated digital-signal processing. Acceleration rates of up to 32.2 ft/sec$^2$ (9.8 m/sec$^2$ or 1 g) are then possible.

However, as long as the heat loss within can be restricted with digital amplifiers and properly sealed magnetic tracks, linear motions' clear advantages lie in the contactless transmission of force, lack of backlash, and much higher response time for acceleration and deceleration.

## Acceleration

Fast rotational spindle speeds of 20,000–30,000 rpm are meaningless for high-speed machining unless they occur with fast axes movements of up to 64.3 ft/sec$^2$ (19.6 m/sec$^2$ or 2 g) and rapid traverse of about 656 ft/min (200 m/min), which in effect cut the machine's positioning time in half (see Figure 4-4).

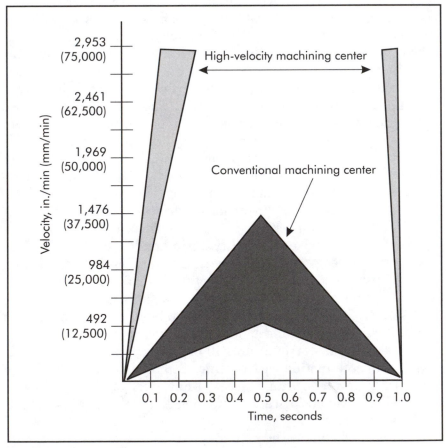

Figure 4-4. Machining centers' axes acceleration.

What holds true for often unnecessary, extreme high spindle speeds beyond 18,000–20,000 rpm within regular part production envelopes is also true for excessively high acceleration rates, as shown in Figure 4-5. Productivity and precision must be paired with economics in regular manufacturing environments. However, there is more to be said about acceleration. Due to its exponential relationship, it takes four times the amount of power to double acceleration. Conversely, deceleration takes four times longer with twice the amount of top speed. The relationship of a machine's top speed to its acceleration is shown in Figure 4-6. For example, a machine with a top speed of 200 ft/min (61 m/min) takes four times as long to accelerate to 50 ft/min (15 m/min) as a machine with a top speed of 50 ft/min (15 m/min). This means that a machine turning at 200 ft/min (61 m/min) moves extremely fast from point $A$ to distant point $B$. However, if the same machine moves from point $A$ to a close point $B$, it takes longer than it does for a machine that moves at 50 ft/min (15 m/min).

Rotational spindle speeds, axes, accelerations, and rapid traverses have to be specified coordinately and tuned to one another according to the respective machining/manufacturing tasks.

## Spindle Lubrication, Cooling, and Chatter

The thermal energy generated during high spindle speeds is substantial, resulting in spindle expansion. The accuracy of the spindle is directly related to stabilizing it and minimizing bearing growths. Oil/air mists lubricate and cool spindle ball bearings best. Their applied volume depends on spindle horsepower and required rotational speeds.

Every precision machine tool has to be perfectly balanced according to the required balance grade. Any imbalance, particularly at high speed, translates into the cutting tool superimposing on its respective imbalance. Machine tools have to be optimized to suit the requirements of the best spindle speed. This is calculated from the frequency response function of the machining operation and the cutting tool. There are systems that use built-in microphones for listening to the cutting operation to identify possible chatter frequencies. In the case of chatter, the system automatically determines the optimum spindle speed by comparing the chatter frequency with the tool pass frequency. Using the "best

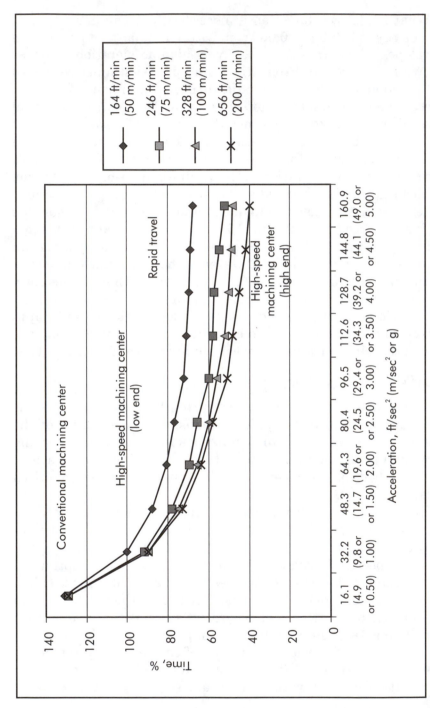

Figure 4-5. Positioning time relative to acceleration. (Courtesy Creative Evolution)

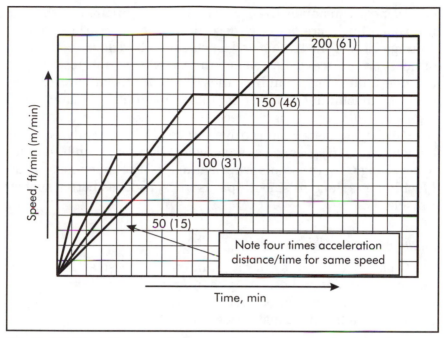

*Figure 4-6. The influence of top speed relative to acceleration.* (Courtesy Creative Evolution)

speed" method optimizes part-surface finishes and cutting-tool life by providing even material-removal rates and minimized vibrations. Making the best speed a machine tool's top speed may indeed determine where the machines' ideal high spindle speeds lie (Schulz 1996).

## Control System

For high-performance machining, computer numerically controlled machine tools have to be very powerful, fast, responsive, flexible, and communicative in both hardware electronics and software programming.

### Interpolation/NURBS

The direction given to the machine is dependent upon the part to be machined. For prismatic workpieces, dealing with straight lines and radii, circular interpolation should be used. A powerful

CNC control can handle the new wave of milling diameters through interpolation, incrementally describing a circle rather than to "plunge" it. On the tooling end, such controls reduce the number of bore- or surface-dedicated cutting tools. It is not necessary to use non-uniform rational B-splines (NURBS) to cut circles. However, for machining complex contours, including arc and linear segments without changing deceleration and acceleration rates, NURBS is the better alternative. Since NURBS really is a continuous curve in space, meaning acceleration rates can be defined at any stage, there is no abrupt change of the machine's mode. Machining is done in smooth paths and produces better surface finishes.

## Intelligent Controllers

PC-based controllers offer a common CNC base and operating platform. Besides orchestrating regular machining functions, the control system of a high-performance machining center must also feature the following capabilities:

- Process diagnostics—including remote diagnostics from the manufacturer to the end-user to monitor and correct machining results;
- connectivity—letting users integrate systems with the machine control (ethernet for example);
- conversational programming—the kind of operator 3D programming that eliminates office support;
- feature recognition—knowledge-based machining software that bases machining sequences on previous experience;
- software library—where the software recognizes and analyzes automatically the geometric paths of machine and tool and determines the area where it has to be segmented; and
- open architecture—this is a matter of compatibility and standardization among varying machine tools. Instead of using proprietary software, open controllers are needed. High-performance machine tools must be equipped with standardized hardware and open-architecture software packages. End-users can then have the control systems with specific functions customized according to their machining needs.

## Geometric Intelligence

Geometric intelligence provided by the "look-ahead" function is a significant feature of controllers, securing high-performance machining (see Figure 4-7). Combining speed and acceleration with accuracy is the true challenge. Feed rates relative to desired finishes are also part-geometry compensation functions. Geometric intelligence's primary functions are to smooth the machine's motions and forego abrupt and erratic runs and stops during the machining process. The smoother the axes' motions, the better and more controlled the parts' surface finishes and the longer the machines' lifetime. This is enabled by software commands. The "look-ahead" software analyzes the machine characteristics and makes the decision to plan ahead for avoiding undershoots and overshoots during machining of complex parts. The program calculates the optional acceleration/deceleration rates relative to the machine's limits. (With earlier closed-loop systems, the controller tendency was to undershoot, while the machines initially tried to overshoot.)

So, geometric intelligence has essentially two functions, feed control and corner deceleration. Feed-forward control looks ahead

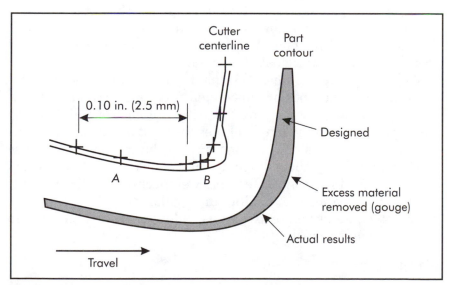

*Figure 4-7. Look-ahead geometry.*

of the anticipated path and the given feed rate. It calculates the error amount that would occur and quickly compensates for the error before it occurs. The amount of correction is controlled for each machine axis correspondingly. Also, when cutting at high feed rates, the system can optimize the feed and program the path to cut with fine incremental settings. This stabilizes the motor's and axes' forward motion, thereby minimizing vibration. Optimum corner approaches, particularly during tool interpolation to produce bores and complex surfaces, are a matter of the right deceleration and the right time. This means to prevent sudden stops and forego system shock and machine vibration. The smoother the deceleration approach to the past corner, the better the surface finish.

Without geometric intelligence included in the controllers' software, high-performance machining, if process accuracy and precision part finishes are part of the equation, would be impossible to accomplish.

## SUPPORT TECHNOLOGY

Within the realm of machining with demanding parameters and expectations, machines must be equipped with state-of-the-art technology. This is not just for a specific run of parts or a task that is at hand right now. *Agility* is defined as the readiness and preparedness for (anticipated) change.

Meticulous attention has to be paid to the technology that supports the machine's performance such as chip, heat, and dust control. Machine tables and part fixturing are part of the machine envelope. Most production machines for most operations use cutting fluids (coolants). Provisions have to be made for their use and handling. What about dry and near-dry machining? How would the basic machine design differ from wet machining? For example, having the capability of adding an angular head to the spindle gives a machine two more axes, thus adding another dimension to the performance capabilities.

### Workholding

Part families and part variety demand much more from workpiece holding than clamping on dedicated lines where one part type is machined. The flexibility of using modular fixturing does not just include the ability to conform to variations in shape and form, but also in part material and workpiece stability. Of course, the machining parameters have something to do with it too. High-cutting-speed machining is usually done with shallow cuts, often in nonferrous metals with relatively low tangential forces. Here, lighter clamping in terms of design and force is appropriate. Many workpiece designs leave the cast part with only thin walls and a hollow shell to be finish machined fast and accurately. Only slight pressures to tie the part onto the table can distort such fragile designs.

The situation is different when cutting with high feed rates in cast iron. With more forces working on the part, a very rigid clamping system secures a firm and precise hold. Chucking parts as few times as possible should be a basic rule. Otherwise, there could be too many variations, especially when several machining centers perform the same operation in a parallel setup. To avoid deformation of delicate parts, such as connecting rods and cylinder pistons, they must be held in position through form and pressure. For example, some valves and sleeves have a tendency to spring back after machining. This has to be considered when specifying the right fixture.

Any fixture design must be such that it permits the greatest number of operations to be performed in one setup. Table arrangements, on which the fixtures are mounted, vary with the family of parts. The machine may be self-contained for multi-task machining or connected to another via a shuttle system. Connecting to another machine would add flexibility for loading and unloading and would bridge gaps in machining time due to tool changes, as shown in Figure 4-8 (ready-made table). For example, a machine with pallet-change times within 5-10 seconds has multi-position, dial-index tables with directional rotation, providing a rotary axis for the machining center. For simultaneous five-axis machining, these rotary tables must deliver high speed, high torque, and extreme accurate positioning (error-free indexing) of five arc sec. and repeatability of ± two arc sec. Direct-driven rotary tables with speeds up to 300–400 rpm can produce perfect circles with excellent surface finishes.

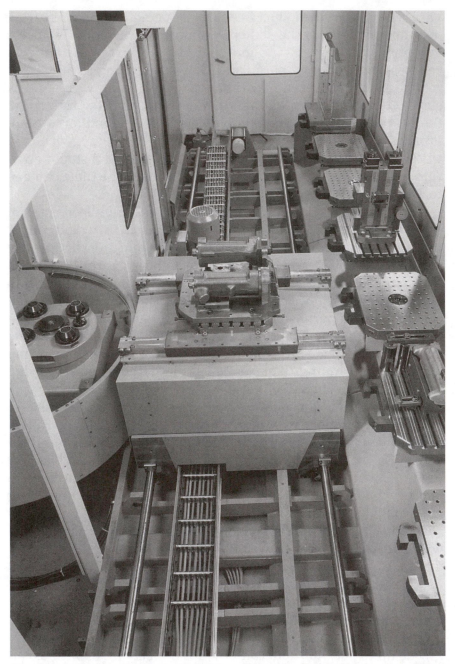

*Figure 4-8. Workholding shuttle tables.* (Courtesy Deckel Maho Geretsried GmbH)

## Dual-spindle, Multi-axis turret

Tool changing and indexing are becoming faster. It is common to see a fast tool changer with tool-to-tool time of about 2 seconds and chip-to-chip time from 4–5 seconds. However, in multi-tasking, high-speed machining, this idle time may still be unacceptable. The problem can be further aggravated as the machine is running at high rpm. It takes time for the spindle to accelerate and decelerate for each tool change. Manufacturers are building dual-spindle machines; one spindle machines while the other is in tool-changing mode to eliminate tool-changing time.

Another popular alternative is the multi-turret head. If the goal is to continuously cut varying parts with precision and productivity limitations not withstanding (high initial cost, tool changes mandating pause of production), multi-axis turrets can deliver quite a punch in performance and precision. Dial-index (rotary), linear-transfer-type systems and machining centers can be equipped with multi-axis turret modules for low- to high-volume production at high cutting speeds. Advanced multi-axis (four and five axes) machining centers can completely finish machine an entire part with several tools cutting simultaneously. This is done in a fraction of the time it takes for a single- or dual-spindle machine to phase-in the needed tools one by one. Tool changing does not require access to storage and the magazine through a tool-change system. Therefore, the cutting head of the tool turret (two to eight spindle stations per turret) can be positioned to any orientation, foregoing complex rotary axes for complex contouring.

Dual-spindle and multi-axes turrets add another dimension to high performance on the production floor, providing high throughput and high machine uptime.

## Wet and Dry Processes

Throughout metalworking, cutting fluids are prevalent. Both non-geometrically defined tools, such as grinding, honing, etc., and geometrically defined cutting tools may be applied with coolants. Those that are not usually perform non-part penetrating operations (some milling and turning, depending on the part material). Acquisition, maintenance, and disposal of coolants are expensive. Coolants also pose hazards for the environment and worker health. On the other side of the spectrum, high-quality part finishes and high machining and manufacturing productivity can only be achieved through the use of coolants. Cutting tools and machine tools, in general, are maintained through coolant agents. Most reputable coolant manufacturers deliver products of equal value as far as the chemical compositions are concerned and can fine-tune their mixtures according to the application.

The most important aspect, from a production point of view, is the proper maintenance of the coolant. This means it must maintain its clarity, be clean for lubrication, and be able to cool the machining process.

Providing chip discharge is a function of coolant pressure to flush the chips away from the immediate work area and carry them out of the machine (conveyors). However, there is also the significant aspect of handling chips that is frequently overlooked. The amount of chips from high-production output and machining with high feeds and speeds and continuous three-shift part runs is directly related to these parameters. Coolant filtration systems must prevent sub-par or sub-optimum machining results. Small chip particles, called *fines,* lodge between the cutting tool and workpiece and act like a grinding agent. The fines become smaller and smaller as they are recirculated in the coolant system and get more difficult to control. Fines can destroy surface finishes and tight geometric tolerances will be lost. Advanced machining centers must feature coolant filtration systems that filter all particles and impurities down to 39.4 $\mu$in. (1 $\mu$m). Some manufacturers' rule-of-thumb is to filter all fines larger than 10% of the tightest tolerance of the part machined. Proper fine filtering must be part of the overall coolant management strategy.

Machining without coolants, involving operations that are normally done with cutting fluids, has consequences for the machine specifications. However, the switch over from wet to dry involves relatively few technical changes, compared to the otherwise complex machine-tool design.

The most influential factors for machining are thermal stress and chip control. Figure 4-9 shows the design criteria for dry machining. Dry machining equipment must be cooled without dust, chip particles, and mist contaminating sensitive parts such as electronics, drives, valves, etc. As for high-speed machining,

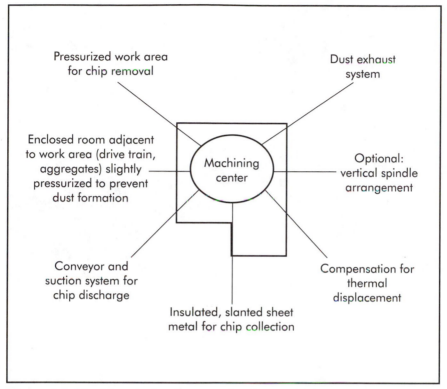

Figure 4-9. Machine design features for dry machining.

the chip volume per time unit increases. In this case, slant bed design and pressurized air consistently remove chips from the immediate cutting area. However, with near-dry machining, the machine tool needs a different supply of minimum volume lubrication fluid through the spindle to provide cooling and lubrication to the workpiece and tool. Thermal stress is still an issue. Therefore, the same design characteristics apply to near-dry as they do to dry machining.

## REFERENCE

Schulz, Herbert. 1996. *High-speed Machining*. Munich, Germany: Carl Hanser Publication Company.

# Advanced Cutting Tools

# 5

The material removal sector of manufacturing is worth more than $40 billion a year, of which about $10 billion are cutting tools, as shown in Figure 5-1. In the U.S., the yearly cutting-tool consumption is about $2 billion. This number has been steady despite the shift of U.S. manufacturing overseas, especially to Asia, Mexico, and Brazil. This transfer, moving the manufacture of parts close to where the end products are purchased, is made possible due to uniform quality standards, the dominance of flexible machines, and common, precision machining processes.

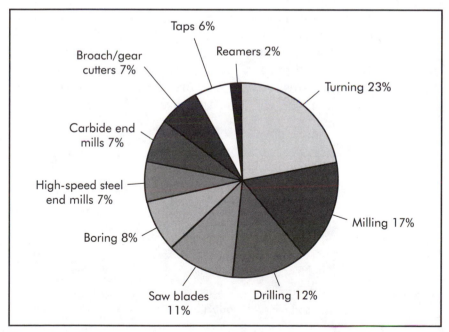

*Figure 5-1. World cutting-tool market* (Courtesy Cutting Tool Engineering).

## TRENDS

The cutting-tool market is very diverse and there are several noteworthy trends. First, the strong shift from high-speed steel to carbide will continue. Second, the shift from carbide to polycrystal-line-diamond cutting material for metal machining has been rather dramatic. Third, technological advancements in coatings for almost all traditional cutting tool materials prolong tool life dramatically. Fourth, hard machining using ceramics and cubic-boron nitride is making headway as a substitute for grinding operations.

Users are demanding tooling that can complement machine-tool capabilities. They want cutting tools designed for high cutting speeds and feed rates. These same tools also need to perform well for lower speeds and feeds when appropriate. Cutting-tool technology must address the public's demand for a cleaner environment, better health, and machined parts finished in fewer hours than before.

Manufacturing companies are seeking to increase productivity and profitability. Some cutting-tool manufacturers do not seem to be in a position to compete in a rather demanding industry and seek to consolidate with other, often competitive enterprises. Outsourcing, seen by many as the only means to compete techno-logically and price wise, has the metalworking market more dispersed than before. In addition, cutting-tool manufacturers often become full-service suppliers—the single source for an original equipment manufacturer 's (OEM's) production floor (see Figure 5-2). This, if done right, can lead to more productivity improvement and cost savings.

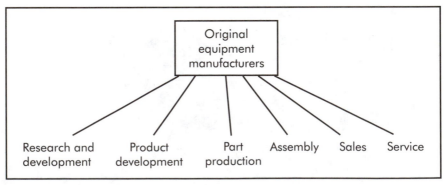

*Figure 5-2. Super-contractor suppliers.*

Subcontractors are held liable by the original equipment manufacturer for holding exceptional part finishes. For the cutting-tool manufacturer, this translates into offering attractive products that consistently achieve extraordinary quality.

It is not good enough to offer a single improvement of one part of a cutting tool. The requirements are for more complete tooling systems. Only they can tackle the more complex and stringent demands of processes such as high-speed machining, near-dry machining, and one-pass machining.

So, for the cutting-tool manufacturer, the guarantee for future success and growth is to be innovative, technology-oriented, and establish machining methods and systems that constitute benchmarks in their field. To achieve that goal, the process of *benchmarking* is done first. This process is a comparative measure against industry, market leaders, and the competition (see Table 5-1). It offers the following opportunities:

- exploring industry's best practices;
- concrete understanding of competitors;
- new ideas for product, process, and practice;
- objective evaluation, reality of company's performance;
- getting an inside view of "us and them";
- uncovering problem areas and opening up opportunities;
- a comparative measure against others in cost, quality, and time; and
- defines the changes necessary to close or bridge possible gaps.

Thus, benchmarking can help in the decision of what cutting-tool systems to pursue.

Table 5-1. Benefits of product and process measures

| Product | Process | Benefits |
|---|---|---|
| Performance | Cycle time | Material flow |
| Function | Throughput | Just-in-time delivery |
| Quality | Machining steps | "Lean" principle |
| Life span | Part transportation | Supply-chain efficiency |
| Reliability | Inspection | Delivery integrity |

Ultimately, the cutting-tool manufacturer that offers productivity, simplicity, and precision at a favorable price and performance ratio, will be the tooling expert in its field against which others benchmark. Maintaining a leadership position will then greatly depend on continuous-improvement efforts and continued technological leadership. A leader in cutting-tool technology must have intricate knowledge of every individual tooling module and characteristic, particularly cutting materials, cutting geometry, special design features, toolholding, and finishing requirements.

## CUTTING-TOOL MATERIALS

The challenges posed by the chemical and physical composition of workpiece material and the capabilities of machine tools have to be met by choice of the appropriate cutting-tool material. Machine tool and cutting technologies must complement each other to reach unheard of productivity gains and part finishes along with more economical use.

Over the years, especially during the past decade, applicable machining data have undergone dramatic changes, which resulted in greater selections of many cutting materials. Improving upon existing cutting material, cutting-tool manufacturers can only alternate within the parameters of more or less hardness and toughness—two opposing material characteristics, as shown in Figure 5-3.

High-performance machining can be performed with any of the dominant cutting tool materials, not just the ones most likely used for high cutting speeds and high feed rates, as shown in Figure 5-4. High-performance cutting can accommodate a variety of technical and technological characteristics. The following cutting materials play important roles.

### Carbides

Generally, carbide is the most versatile and widely used cutting-tool material in chip-making manufacturing. In fact, all other cutting-tool materials are compared to the performance of carbide. Cemented carbide is composed of tungsten and carbide powder,

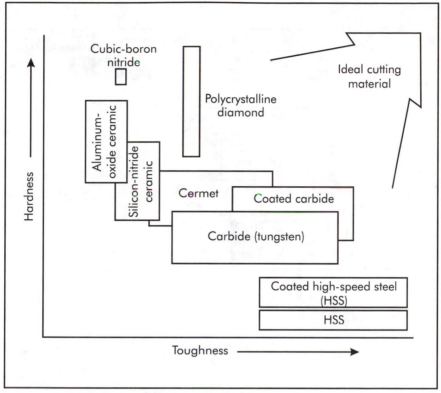

*Figure 5-3. Hardness and toughness of various materials.*

mixed at a temperature of about 3,400° F (1,871° C) and carbon-ized to tungsten carbide. Depending on what characteristic is de-sired, other additives are cobalt, niobium, titanium, and tantalum. Then, the powder mix is compacted and subsequently baked (sin-tered at a temperature of about 3,000° F (1,649° C). The end prod-uct is a precise, chemically defined carbide grade.

There are many different grades for a wide range of operations. For example, by definition, grades C-1 through C-4 contain car-bide and cobalt binders, while grades C-5 through C-8 contain titanium and tantalum carbide. In general, higher-grade num-bers denote greater tool hardness, while lower-grade numbers are indicative of more toughness. Lower-numbered grades are more suitable for heavier, roughing cuts, while higher-numbered grades are more wear resistant and for finishing cuts.

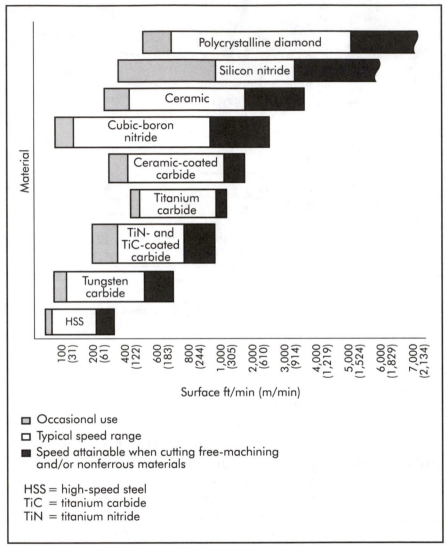

*Figure 5-4. Speed ranges for various cutting-tool materials.*

Some cutting-tool manufacturers have been experimenting with the size of the individual powder particles and have developed a "micrograin" carbide. Simultaneously increasing the cobalt content achieves a carbide grade that offers better-than-average toughness and good resistance. It is harder than any of the standard C-1 through C-8 carbides. Micrograin carbide tools are effective substitutes for regular carbide tools. They are used for extra-high stock removal and high-abrasive-wear machining. Micrograin C-2 is the most popular carbide on the C scale. This grade, in principle, retains the regular C-2 toughness and offers more wear resistance. It has low cobalt content, resulting in longer tool life due to reduced chipping.

## Coatings

The carbide insert's value in machining is immense and its versatility and performance are greatly enhanced through diverse coatings. They offer prolonged tool life, are economical, can be tailored to specific machining parameters, and yield superior part finishes from turning, boring, drilling, reaming, and milling processes. Applied through physical- and chemical-vapor deposition (PVD and CVD, respectively), the most popular coatings are titanium carbide (TiC), titanium aluminum nitride (TiAlN), and titanium nitride (TiN). A single coating layer has a thickness of between 80–200 mils (2,032–5,080 μm), providing an inert barrier to prevent diffusion of the cobalt from the carbide composition.

Coatings, besides prolonging tool life, also help in improving surfaces. They act like a lubricant and assist in staying within tolerance longer, as shown in Figure 5-5. Combinations of multilayer coatings take their favorable properties even further. With multiple coatings, the first layer is usually TiC, because of its good

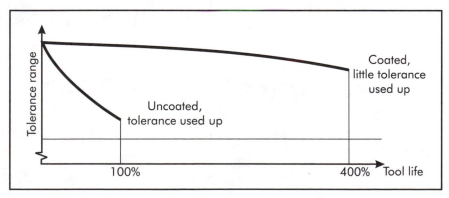

Figure 5-5. Coating versus tool life.

adhesion and thermal expansion to the substrate. The outer layer is selected based on the need for heat, wear, and crater resistance, as well as edge strength and preparation. Outer layers can be TiC or TiAlN, and any of their combinations. The thickness of the layers varies depending on the material cut and the operation. For example, milling calls for a rougher outer layer than finish reaming.

Another variation of applying coatings is partial layering. An example is shown in Figure 5-6, which depicts two cutting inserts with different partial coatings. "Z" is used for finish-machining steel and cast-iron alloys, and "X" for heavier cuts in steel.

Gradient carbides are also coated carbides, the outer layer of which is enriched with cobalt, while its inner structure is identical to other carbides, as shown in Figure 5-7. Since cobalt lends more toughness to the outer layer, the cutting edge is less brittle. Therefore, it withstands mechanical shock better. This is of significance when machining part configurations with interruptions.

Zirconium nitride (ZrN) coatings offer unique comprehensive properties. Their thin, hard layers of more than Vickers hardness

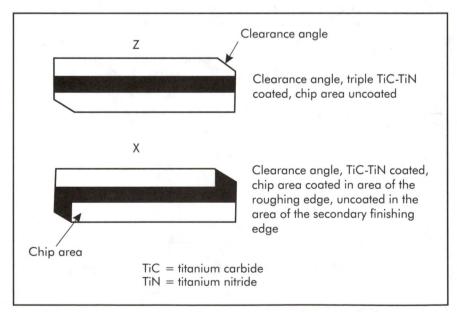

*Figure 5-6. Partial coatings.*

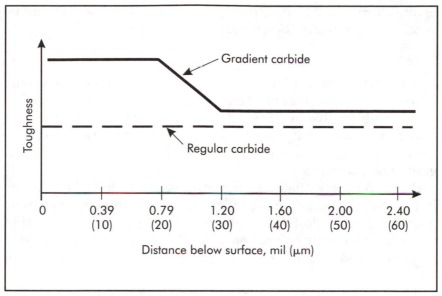

*Figure 5-7. Toughness of gradient carbide.*

(HV) 300 add lubricity, ductility, and oxidation resistance in one layer. Zirconium nitride coating can be used for ferrous and non-ferrous material and is especially suited for drilling, reaming, and milling processes. Titanium boride ($TiB_2$) coatings are being developed especially for machining aluminum because they resist piece-material build-up during cutting.

## Cermets

When carbides containing niobium, tantalum, or molybdenum are added to a titanium nitride base, cermets are formed. Their fracture toughness and heat and wear resistance make them able to withstand high cutting temperatures. They run at high cutting speeds and can be used for machining hard steels and hard cast irons above Rockwell hardness (HRC) 40. Even Inconel® or Hastelloy® can be finish-machined. However, it should be noted that cermets cannot take heavy cuts. When used in conjunction with thin-layer coatings, cermets offer extended tool life. Since their price tag is the same as regular carbide, it makes them an interesting alternative. In fact, they are under-utilized and should be specified more often. (See Figure 5-8.)

## Ceramics

Although proposed for machining in 1905 and patented some 10 years later, ceramics have only more recently been accepted as a viable alternative to other cutting-tool materials. This was due to their non-uniformity and weakness (brittleness). However, they were only used on older machine tools that lacked the needed rigidity and accuracy.

Since the early 1980s, many improvements have been made in the microstructure of ceramics (grain size and density), their processing, and additives available. This has resulted in cutting materials that can be applied at high cutting speeds and still yield extended tool life in milling, drilling, tapping, and turning operations.

The ceramic tools available consist of two base materials, aluminum oxide ($Al_2O_3$) and silicon nitride ($Si_3N_4$). In their purest form, they are very limited for demanding and economically efficient machining, primarily due to their brittleness. Another shortcoming is their low thermal shock factor, which somewhat limits the free use of coolants. See Figure 5-9.

However, the right combination of additives allows ceramics use in specialized, selective applications. They are hard to beat because they resist scaling and crater wear better than tungsten carbides. They also can operate within a broader temperature range, depending on cutting speed and stock removal. This results in better surface finishes and defined chip control. The most prevalent ceramic cutting materials are aluminum oxide with zirconium-oxide additives, whisker-reinforced aluminum oxide, and silicon nitride ($SI_3N_4$).

### Aluminum Oxide with Zirconium-oxide Additives

Adding TiN to $Al_2O_3$ adds 30% more to aluminum oxide's rupture strength, which is still below that of carbides. However, this ceramic can endure much higher speeds. It can be used for general machining of cast irons and steels up to HRC 60 (stainless, alloys, and heat-treated materials). However, interrupted cuts, high feed rates, and high stock removal should be avoided.

---

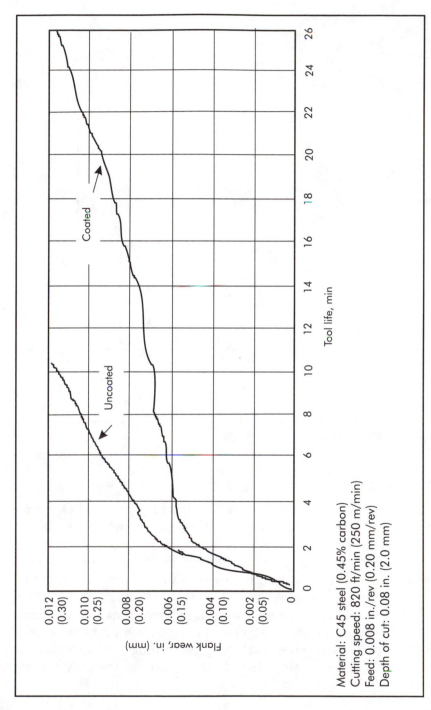

Material: C45 steel (0.45% carbon)
Cutting speed: 820 ft/min (250 m/min)
Feed: 0.008 in./rev (0.20 mm/rev)
Depth of cut: 0.08 in. (2.0 mm)

*Figure 5-8. Tool-life comparison between coated and uncoated cermets.*

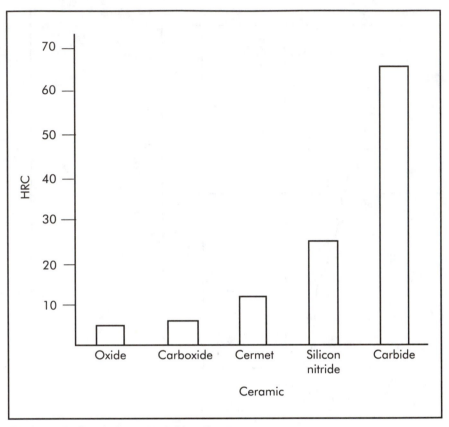

*Figure 5-9. Ceramic material hardness.*

## Whisker-reinforced Aluminum Oxide

The inclusion of silicon carbide (SiC) whiskers in $Al_2O_3$-based ceramics increases its toughness. SiC whiskers, which can be up to 50% of the content, increase the thermal-stress coefficient of the ceramic. Furthermore, the different thermal-expansion coefficients of $Al_2O_3$ and SiC lead to inherent push forces, counteracting the pull forces generated through cutting. The higher toughness combined with high wear resistance make it a suitable cutting material for tough steel and even nickel and chromium-based alloys. Unfavorable machining conditions such as interruptions and high feed rates should be avoided, since they reduce ceramic tool life exponentially.

## Silicon Nitride (SI₃N₄)

Silicon nitride is excellent for machining gray cast iron. Its thermal shock factor is higher than that of any other ceramic. This, combined with high toughness, lets it successfully machine parts with interruptions at high feed rates with regular coolant supply. Silicon nitride does not cut steel well, since the temperatures generated are too high and it breaks down within relatively short machining times. The cutting edge of $Si_3N_4$ loses its sharpness shortly after the first cuts in any material before it stabilizes itself. With a small radius, fine finishing is only advisable if there is an absolutely stable machine-tool fixture setup.

Silicon nitride ceramics, also known as sialons, are superior for cutting gray cast irons, increasing cutting speeds up to 700% over carbides. Considering that silicon is abundantly available as raw material, it surely will be further developed to enlarge its applications.

## Polycrystalline Diamond (PCD) Coating

No other cutting material is experiencing the same growth in use as PCD, especially for high cutting speeds, nonferrous machining, and finish operations of any kind. When diamond powder, sintered to one uniform mass with thickness of approximately 0.02–0.03 in. (0.5–0.7 mm) thickness is pressed onto a cemented carbide substrate, PCD is made. The micrograin structure, varying between 0.177–3.900 mil (4.50–100.00 μm), gives PCD its physical characteristics. The coarser the better the abrasion and impact resistance, as shown in Figure 5-10.

PCD exhibits superior performance at high cutting speeds, lower feed rates, lighter cuts in nonferrous materials, and particularly in semifinishing and finishing bores and surfaces with tight tolerances and stringent requirements. Tool life increases of 300–400 times that of carbide are possible, as shown in Figure 5-11. Because of PCD's resistance to part-material built-up during cutting, it is also ideally suited for dry- and near-dry machining. PCD-tipped tooling offers superior hardness, resistance, and fatigue strength. However, it needs a very stable, robust, and accurate machining setup and special handling because it can easily get chipped and damaged. If properly applied, PCD can be reground

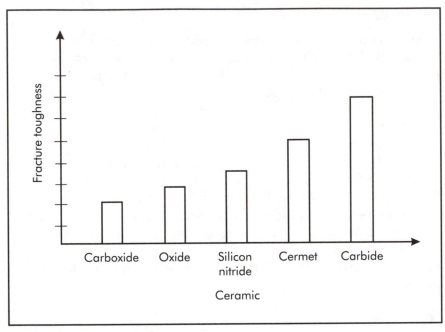

*Figure 5-10. Fracture toughness.*

several times and still yield the same excellent machining results and a very favorable price/performance ratio.

## Diamond Coating

Given the expected, continued exponential growth of the use of nonferrous metals, notably aluminum in the automotive industry, the use of diamond cutting-tool material will be extraordinary. The average price for a simple diamond insert is about four times the price of a comparable carbide insert. Hence the quest for cutting-tool materials with lower purchase prices. Diamond coating, also known as chemical-vapor deposition (CVD) diamond, could possibly be a rather formidable alternative to PCD inserts and PCD-tipped tooling, as shown in Table 5-2. CVD diamond features a dense polycrystalline structure with a tensile strength of 10–17 lbf/in.$^2$ (7,000–12,000 kg/mm$^2$) and a coefficient of friction of 0.05–0.15. It is generally up to 15% harder than normal PCD. Its abrasion resistance is equal to that of PCD.

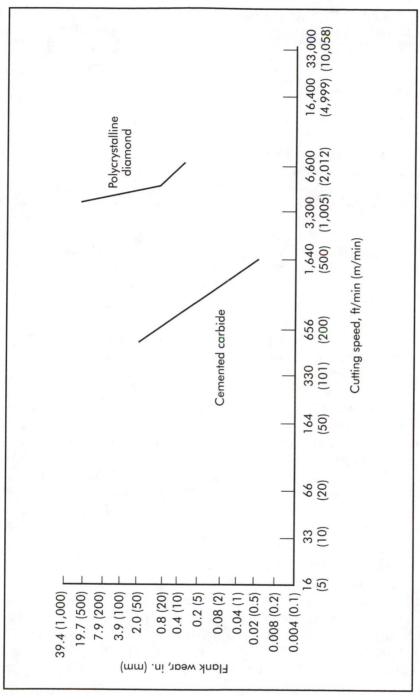

*Figure 5-11. Flank wear and cutting speed for cemented carbide and polycrystalline diamond.*

Table 5-2. Physical characteristics of chemical-vapor-deposition (CVD) diamond and polycrystalline diamond (PCD)

|  | CVD | PCD | Cemented Carbide (C-2) |
|---|---|---|---|
| Density, g/cm$^3$ | 3.51 | 4.10 | 15.00 |
| Young's modulus, GPa | 1,180 | 800 | 600 |
| Compressive strength, GPa | 16.0 | 7.4 | 5.0 |
| Transverse rupture strength, GPa | 1.3 | 1.2 | 1.7 |
| Fracture toughness, MPa/m$^{1/2}$ | 5.5 | 9.0 | 11.0 |
| Knoop hardness, GPa | 85–100 | 50–75 | 18 |

CVD diamond's thermal stability is better than PCD because of its purity in carbon and hydrogen. This means it can withstand very high temperatures without breakdown. Adding to these characteristics of variability and lower cost, diamond coatings clearly come out on top.

The problem is the lack of adhesion and peeling found with CVD diamond. Cobalt, an otherwise desirable ingredient of tungsten carbide, lends toughness to its composition. However, it becomes a catalyst for graphitization during machining when coated with a diamond film, which is an agglomerate of small crystals strongly bonded to one another. The result is poor adhesion to the tungsten carbide substrate, which causes the diamond film to peel off. Nevertheless, cobalt is necessary to lend toughness to the cutting material. Currently, it appears that CVD diamonds with a silicon-nitride substrate, such as $Si_3N_4$, can be successfully used as the substrate for PCD. This may be done only if brittleness during machining is not an issue. Drilling and reaming processes have been successful at lower cutting speeds for materials such as a low-cobalt-tungsten carbide of less than 5%, multi-layered materials, or $Si_3N_4$ substrate, as shown in Figure 5-12.

Diamond coatings have not been adopted for regular machining on demanding production floors. However, there are many studies being done in industry and academia. Some experts believe that the adhesion problem will soon be solved and PVD coatings will play a pronounced role in volume machining.

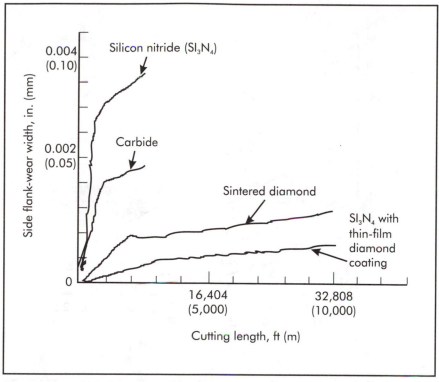

*Figure 5-12. Side-flank-wear width versus cutting length.*

## Cubic-boron Nitride (CBN)

Cubic-boron nitride is produced by the high-temperature, high-pressure sintering (similar to PCD) of CBN particles and a binder material. The raw material is cubic-boron-nitride grit that has been used for grinding abrasive ferrous metals. CBN cutting inserts are mostly dipped, that is, they are bonded to carbide-backing layers as compared to solid, unbacked inserts. In hardness, CBN ranks second only to PCD. It has a low affinity to ferrous metals and remains chemically stable when machining them, even at high cutting speeds. Its mechanical and thermal properties make it a superior cutting-tool material for difficult-to-machine, hard, ferrous alloys. Its hardness at 1,300° F (704° C) is still better than that of carbide or ceramic at room temperature, as can be seen in Figure 5-13.

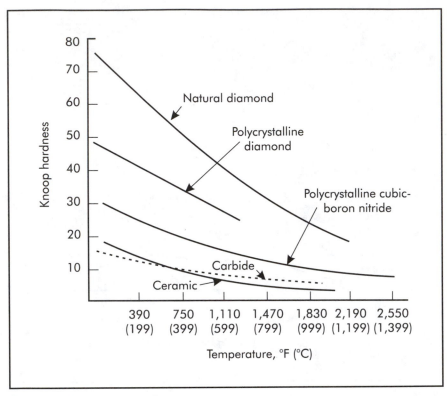

*Figure 5-13. Knoop hardness.*

At high temperature, carbide tools tend to soften at a rate similar to that of the workpiece material. This combined with chemical reactions (oxidation) limits carbide's practical use to temperatures below 1,300° F (704° C). The temperature resistance and conductivity of CBN tools make them highly suited for running at high feed rates and cutting speeds where ceramics and carbides would deteriorate.

CBN was developed to prepare and finish-machine hardened steel, hard cast iron (nodular and malleable), and superferrous alloys with a high content of cobalt and nickel.

Often, parts previously finished by grinding can be more productively and economically machined through turning, milling, and fine boring, depending on the microstructures of the workpiece materials. For example, the machining parameters must be optimized for perlitic or ferritic materials, or both. The same holds true for proper coolant. Depending on the application and the technology at hand, CBN tools can sometimes even run dry if proper chip discharge is provided through pressurized air.

PCD and CBN superabrasive cutting-tool materials for nonferrous and ferrous workpieces are the materials of choice for high-cutting-speed machining. This is despite their higher initial cost when compared to conventional cutting material because of their superior hardness, chemical inertness, and heat conductivity. PCD and CBN cutting-tool materials guarantee robust machining with extreme machining data, without running the risk of premature tool failure and damaging and deforming the workpiece due to thermal stress. Price/performance ratios, the number of parts to be machined, the machine tool available, the selected machining process, required part finishes, etc., all play a part in deciding which cutting-tool material to use. For high cutting speeds, Figures 5-14 through 5-18 compare suitable cutting-tool materials for specific workpiece materials.

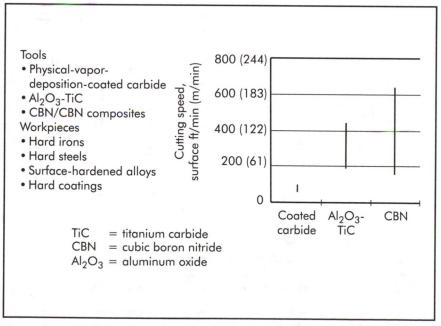

Figure 5-14. Tool vs. workpiece materials for high-speed hard machining.

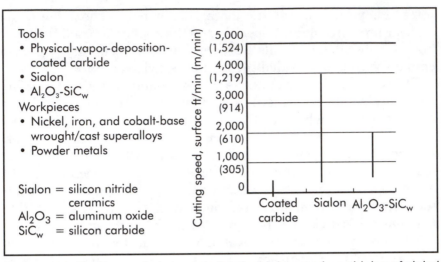

*Figure 5-15. Tool vs. workpiece materials for high-speed machining of nickel alloys.*

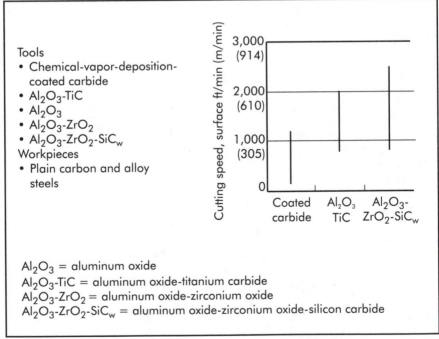

*Figure 5-16. Tool vs. workpiece materials for high-speed machining of steels.*

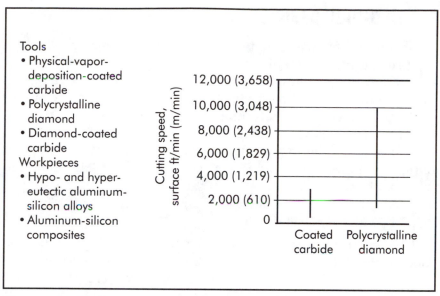

Tools
- Physical-vapor-deposition-coated carbide
- Polycrystalline diamond
- Diamond-coated carbide

Workpieces
- Hypo- and hyper-eutectic aluminum-silicon alloys
- Aluminum-silicon composites

*Figure 5-17. Tool vs. workpiece materials for high-speed machining of aluminum-silicon alloys.*

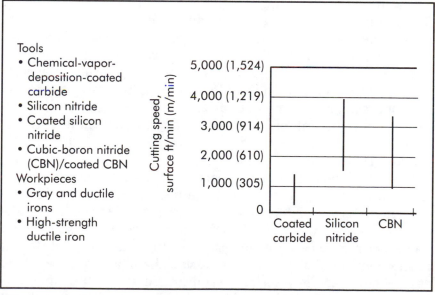

Tools
- Chemical-vapor-deposition-coated carbide
- Silicon nitride
- Coated silicon nitride
- Cubic-boron nitride (CBN)/coated CBN

Workpieces
- Gray and ductile irons
- High-strength ductile iron

*Figure 5-18. Tool vs. workpiece materials for high-speed machining of cast iron.*

# CUTTING-TOOL DESIGN

Workpiece shape and material, as well as the selected machining process, are the main determinants of cutting tools. The success of any chip-making process is determined by what happens at the point of contact between workpiece and tool. The geometry of the cutting edge is of pivotal importance to machining, in addition to other intricate elements that make the perfect high-performance tool. The important criteria are: geometry, machining data, tool guiding, chip and burr control, ease of handling, and adjustability.

## Geometry and Machining Data

The geometry of the cutting tool must be designed to accommodate the demands of higher and, in some cases, extreme cutting speeds, feed rates, higher part-finish requirements, coolant issues, and the elimination of roughing or post-machining processes. These requirements really push the envelope to the limit of current capabilities.

In the turning process, tools with inserts are the best option. The same is true for drilling and boring operations. However, for fine boring and reaming, there are different phenomena. For nonferrous metals, many users prefer indexable inserts because they allow fine-tuning according to the desired blueprint tolerance. When it comes to nonferrous metals, machining with PCD inserts is as common as it is with fixed PCD tooling. It really depends on the application and sometimes on the user's preference. Since inserted tooling has all but conquered advanced machining, this is what the following discussion will concentrate on, with a few exceptions that are noteworthy.

Because of near-net-shaped parts and the increase in nonferrous material use, lighter cuts have become the norm. This, in turn, has also allowed finish machining in one or two passes. The cutting-edge geometry is as shown in Figure 5-19. The primary and secondary cutting edges can be combinations of different angles depending on the part material, the machining data, and desired surface. They perform semi-finishing and finishing during the same cut. Typical for advanced tooling are clamping grooves that secure safe clamping through form and pressure in radial

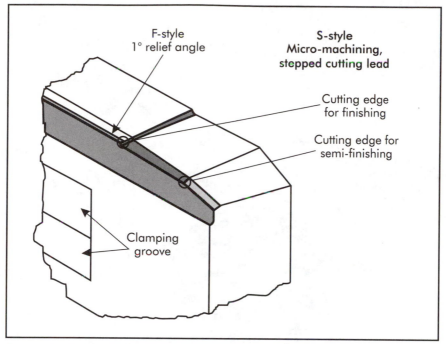

F-style
1° relief angle

S-style
Micro-machining,
stepped cutting lead

Cutting edge
for finishing

Cutting edge for
semi-finishing

Clamping
groove

*Figure 5-19. Stepped cutting geometry.*

and axial directions, enhanced by a rough-sintered surface, as shown in Figure 5-20.

An important third geometry is that of ground-in chip breakers. They are arranged radially and axially aiding to break the chips, making them manageable, and preventing long, stringy chips, which can cause severe damage to the tool and workpiece (scratches and heat build up). Another geometric design feature of the tool is a large chip area that starts with the cutting insert. It is sunken in the tool body with no hardware protrusion to prevent chips from being caught. This ensures smooth chip flow and discharge, as shown in Figure 5-21.

Cutting-tool geometry is directly related to the machining data applied. Which, in turn, has to comply with the performance of the machine tool, the machining process, and the physical and chemical make-up of the workpiece material. It is necessary to find the optimum feed rates and cutting speeds—a balance weighed between production cost, time, and part finish. Higher cutting

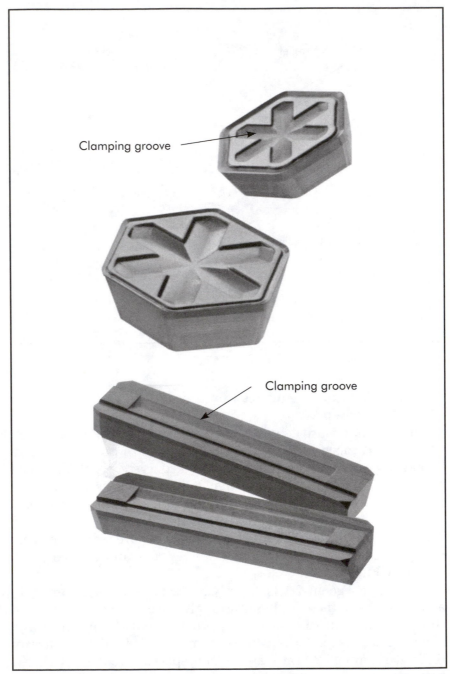

*Figure 5-20. Inserts for form and pressure clamping.* (Courtesy MAPAL, Inc.)

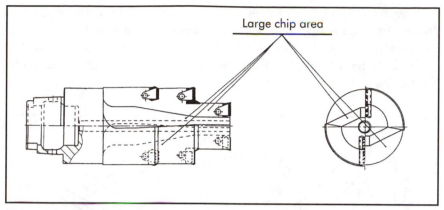

*Figure 5-21. Precision tool for heavy stock removal.*

speeds increase productivity by lowering main machining time, thus increasing output and lowering non-machining time. But this can also lower tool life and increase indirect tooling costs (for tool management, more frequent setting, etc.). The objective is to find the optimum cutting speed for the particular machining process.

## Tool Failure

Tool failures can be caused by disintegration of the tool body itself, lifting of cutter clamp plates, bursting of the cutting inserts, or deformation of connecting hardware (cassettes, bolts, etc.). Design layouts have to be based on empirical tests to determine the elevated speeds at which destruction of the tool begins and where it eventually breaks apart.

## Design Considerations

Design objectives include to:

- Minimize the tool's mass.
- Strive for tool symmetry.
- Build the tool with smooth surfaces.
- Use a minimum amount of hardware.
- Provide form and pressure clamping for inserts.
- Eliminate frequent tool adjustments.
- Smaller elements should be designed to fail first to lower the impact in case of tool failure.

Increased feed rates lower the time the tool is in the cut, thus lowering machining time. But they also increase tooling cost, because the stress on the tool increases exponentially. Again, the right balance of cost increase versus productivity gain has to be found, as shown in Figure 5-22.

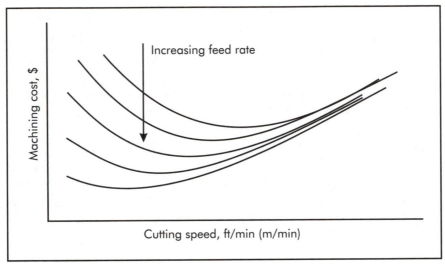

*Figure 5-22. Machining cost vs. cutting speed and feed rate.*

## TOOL GUIDANCE AND STABILITY

As the cutting speed increases, consideration has to be given to balancing and safety. At high speeds, centrifugal forces increase as the square to rotational velocity. Balance of the tooling system becomes an issue. Excessive imbalance of the tooling system has an effect on the workpiece (chatter marks and out-of-tolerance conditions) and can cause premature spindle failure. In worst-case scenarios, imbalance causes a tooling system to cut out of control and break or damage workpieces and presents a safety hazard for machine operators. To avoid such conditions, the toolholder and machine need to be as close to the same balance grade as possible. However, good balancing notwithstanding, sheer centrifugal forces are cause for concern in high-speed machining. Cutting tools must be designed to secure safe operations.

Almost 70% of chip making involves hole making, which is mainly accomplished by precision machining. To arrive at excellent bore geometry, the way the tool is guided through the bore is very important. In the past, bushing carriers, located between the workpiece and machine spindle, supported the tool into the cut. Then, the tool machined the inside of the part. Now, cutting-tool use is shifting to machining centers. Bushings are mostly a thing of the past because of much improved spindle accuracy and advances in tool designs.

In machining centers, cutting forces are absorbed by guide pads, which reduce vibration that impairs the cutting operation. Vibrations or tool chatter can yield unacceptable surface finishes, dimensional workpiece inaccuracies, or premature failure of spindle bearings, producing workpiece scrap.

Besides stabilizing the tool during cutting, guide pads open up new avenues to the insert design. Due to the tool's rigidity, more intricate cutting-blade (insert) geometries can be fitted. For example, smaller radii at the tip and primary and secondary cutting edges can be made with better finishes. Tools can run at higher speeds and feed rates when entering the workpiece, and during cutting, because the tool stabilizes itself as it enters the bore.

Leapfrogging technology often stimulates other unrelated areas, offering solutions to other indirectly related problems. Here too, the separation of the geometrically defined guiding and cutting functions offers answers to otherwise unresolved interrupted cuts, bridging air gaps between workpiece materials on the same centerline, and the undesirable following of previously out of line machining. Bridging air gaps through extended guide pads and overcoming interruptions in bores by providing additional, peripherally arranged guide pads, ensures high-precision machining. Because of rigidity, the tools do not follow the previous path of operation, but rather cut their own and thus automatically correct out-of-centerline conditions during the finishing cut. An important prerequisite for using padded cutting tools is a clean 6–8% coolant emulsion when using carbide guide pads and cermets guide pads for nodular iron and steel.

A breakthrough technology is the design of tools with PCD guide pads. Not only are they not prone to chip build up, but they can be used for dry and near-dry machining and prolong tool life dramatically. PCD-padded tooling makes semi-finishing and fin-ishing of critical bores robust processes.

Contrary to widespread assumption, built-in guide pads do not limit cutting speeds more than any other design. The limitation is inherently in the feed rate as long as there is only one cutting insert. To substantially increase the feed rate, it is necessary to design single-blade, fine-boring tools featuring guide pads with an additional insert (Figure 5-23). This method could be called micro-machining (see Figure 5-24). The two inserts are offset to each other radially so that the finishing insert only cuts a few hundredths of an inch (millimeter), while the roughing path takes

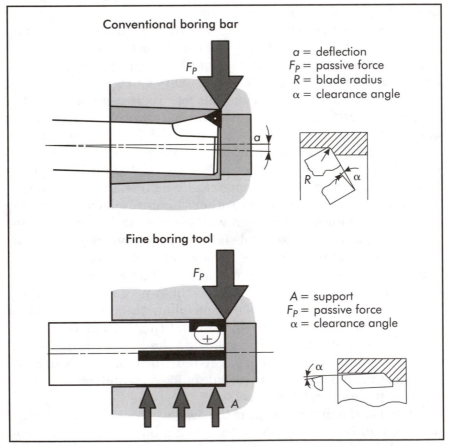

*Figure 5-23. Tool deflection.*

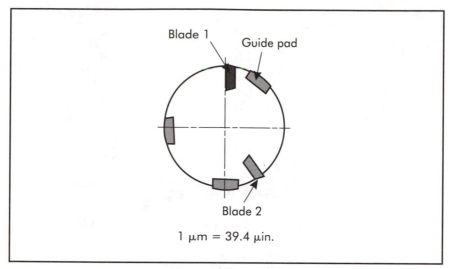

Figure 5-24. Twin-blade design for micro-machining.

out the overwhelming depth of cut. The separation of roughing and finishing cuts secures high cutting speeds and feed rates. As a result of the minimal final touch performed by the finishing insert, the surface finish and tool life can be improved—truly a high-performance process.

Not every finishing tool for boring features guide pads. However, it has been proven in high-production manufacturing that for precision machining, including high cutting speeds, guide pads are of great advantage even in cases of solid, fixed tooling. It is important to stabilize the cut. Another design that can help is helically shaped milling tools, where the inherent shape of the cutter sequentially machines by passing the cut from one edge to the next. Stabilizing cuts can take the aggressiveness out of the cutting edge through additional edge prepping (honing and rounding off).

## Stable Cuts

An often-overlooked aspect is that of the tool body itself compared to the inherent vibration of cutting tools during machining as shown in Figure 5-25. Minimizing such vibrations can be accomplished with tool bodies made of heavy metal. However, securing stable, manageable machining at elevated speeds and feeds

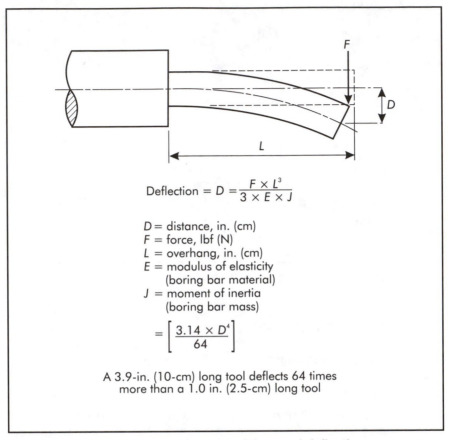

$$\text{Deflection} = D = \frac{F \times L^3}{3 \times E \times J}$$

$D =$ distance, in. (cm)
$F =$ force, lbf (N)
$L =$ overhang, in. (cm)
$E =$ modulus of elasticity
    (boring bar material)
$J =$ moment of inertia
    (boring bar mass)

$$= \left[ \frac{3.14 \times D^4}{64} \right]$$

A 3.9-in. (10-cm) long tool deflects 64 times
more than a 1.0 in. (2.5-cm) long tool

*Figure 5-25. Influence of tool bar material on tool deflection.*

with flexible machining centers is difficult with heavy or relatively big cutting tools. Handling and safety can be tagged-on challenges. Hollow and/or aluminum or magnesium tool bars can remedy the problem associated with the centrifugal force during high-speed cutting. They eliminate the issue of weight limitations and offer ease of handling. Surface-hardened, nonferrous toolbars feature steel cassettes as built-in cutting inserts and, if necessary, guide pads.

Weight mass and centrifugal force, two significant factors of high-cutting-speed machining, have resulted in the explosive growth of face milling cutters with aluminum bodies.

# CHIP CONTROL

Of the machining parameters, feed rates are mostly effective to control the flow of chips. At high feed rates, more pronounced material direction and deflection make the chips break easier and to the desired size. Higher cutting speed and higher feed rate increase the temperature at the cutting point, softening the part material to aid in cutting more efficiently.

Chip making is a by-product of machining, which has to be controlled, particularly in precision and high-performance machining. It is a result of material deformation or plastic flow of material. The criteria surrounding it involves the properties of the workpiece material, machining data applied, and cutting temperature, which affect the thickness and length of the chips and their discharge out of the machining area. The objectives are to:

- generate a smooth flow of the chip along and past the tool face so as to not interrupt its continuous flow to carry with it a big part of the thermal stress; and
- prevent a pile-up or build-up within the chip itself.

Speeds, feeds, and cutting-tool geometry (including chip breakers) have a lot to do with controlling chip formation. In addition, the workpiece material and coolant can be of great assistance in controlling chips. Brittle materials produce short, discontinuous chips. Breaking them at the right point and then properly disposing of them is important. However, ductile materials produce continuous chips, making chip-breaking-insert geometry necessary. It is also prudent at the design stage to specify material properties that can inherently assist in breaking up longer chips. For example, additional amounts of phosphorus, lead, and/or sulfur added to low-carbon steels assist in breaking up longer chips. Coolant can manipulate the flow of chips and their discharge a great deal. It does not suffice to provide coolant volume; what is needed is high-pressure coolant (or air of between 250–500 psi [17.6–35.2 kgf/cm$^2$]). However, the pressurized coolant has to be precisely directed to the tool's cutting edge through defined through-the-tool coolant passages. The coolant must exit the tool hole to cool the process. The hole must be placed where it is needed to break the chips and discharge them outside the immediate machining area. Without proper coolant passages, advanced machining is not possible.

With the trend for smaller workpieces and, hence smaller bores and thinner part walls, extremely high coolant pressure has to be avoided because this can distort the tool and result in out-of-specification parts. Extensive empirical tests involving drilling and reaming tools show that exit geometry is important relative to the performance of the tooling.

# BURR CONTROL

Burrs are a phenomenon. They are undesirable because they negatively affect the proper functioning of parts due to edge interference and obstruction, and can interfere with subsequent parts assembly. For example, burrs are detrimental to the successful use of hydraulics. They also can be a health hazard to workers who handle parts.

It is not so much that deburring poses technical and technological problems. The issue is the cost for subsequent, separate deburring processes to meet part specifications. Efforts are underway in academia and industry to contain the formation of burrs and secure their less costly removal. The leading consortium on deburring is the Laboratory of Manufacturing Automation at the University of California-Berkley (Dornfeld 2001; Hwang 2001). It was established to address:

- problems related to predicting and modeling burr and edge breakout;
- the development of a database (or knowledge base) of best burr avoidance and removal techniques in industry;
- development of a computer-aided design (CAD) advisor—a burr expert for designers and process planners;
- development of strategies for deburring; and
- identification and development of advanced deburring technologies.

Although all manufacturing processes are of interest in the program, of special interest are those processes for precision manufacturing. The products of the consortium are:

- software (CAD Burr Expert®);
- database of burr minimization and deburring knowledge;
- burr-formation models;
- application-oriented solutions to deburring and edge-finishing problems;
- deburring hardware strategy evaluation;
- deburring inspection and burr-measurement technology;
- standard burr terminology and specifications for burr/edge characterization; and
- engineers trained in burr minimization and deburring techniques.

As the studies spill over to industry, manufacturing will get a better handle of how to deal with burrs. Meanwhile, cutting-tool manufacturers' must use built-in geometry to automatically deburr on regular production machines with the same setup for all other operations. One such example is an aluminum milling head, fitted with cartridges, which also incorporates brushes for deburring, as shown in Figure 5-26. After completing the milling process, the milling cutter is retracted. Then, brushes, which are peripherally arranged and built into the milling head, move outward and across the workpiece surface in the opposite direction of the previous milling operation. After the surface is deburred, the brushes are retracted. This cutting tool can be easily adapted to any machine and incorporated in any machining process. It saves by eliminating the need for additional tooling, tool changes, or even machine stations. With robust technologies like this, deburring will be just an automatic, built-in, operation.

## TOOLING ADJUSTMENTS

For many applications, end-users prefer cutting tools that do not have to be preset. They demand tooling with solid cutting edges, and inserted tools, where the cutting insert rests in a pre-machined pocket to be tightened by only one screw.

Manufacturers of precision cutting tools have CNC machines, including grinding operations with high accuracy to produce advanced fixed tooling. They can finish-machine PCD-interpolation tools, on which even international tolerance bands (IT5/IT6) can

*Figure 5-26. Precision milling head with cartridges.* (Courtesy MAPAL, Inc.)

be held. The advantage of precision solid tools is that they can be taken out of storage and put in the spindle without setting.

Other tools require setting. They feature hardware that sets tools according to the blueprint diameter and to cut within the respective tolerance band. In between the two extremes are the compromise tools that offer adjustments and fixed inserts, as shown in Figure 5-27. Slots provide for precise seating of the cartridges. Adjustment of the cartridge only in the axial direction guarantees the lowest possible axial runout. High-tensile-strength mounting screws position the cartridge accurately and with high rigidity. The axial adjustment is done easily and quickly.

Cost-cutting efforts invite standardization as much as possible within the entire envelope of manufacturing and machining. In chip

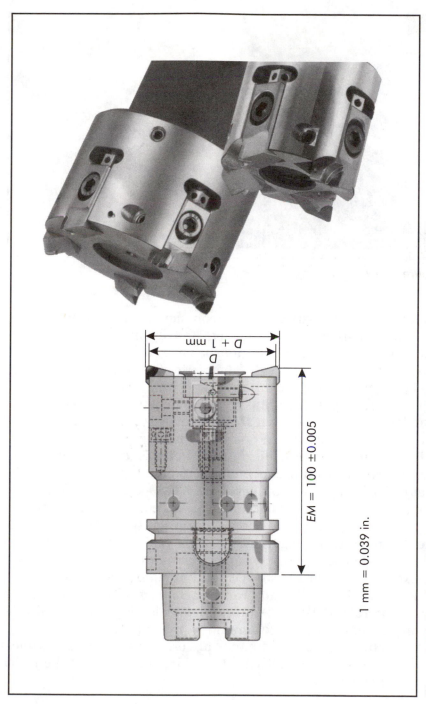

Figure 5-27. Precision ISO inserts. (Courtesy MAPAL, Inc.)

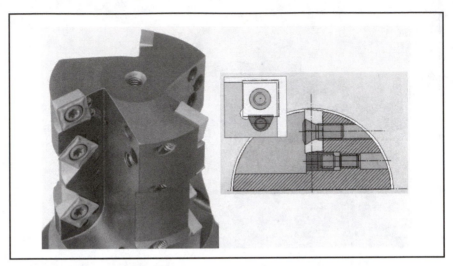

*Figure 5-27. (continued)*

making, ISO standard inserts have conquered almost all operations, especially turning, boring, grooving, and drilling. Because of standardization, such inserts are inexpensive to purchase and users are familiar with their shapes. Their ease of handling and adequate performance make them desirable and well accepted.

Before high-precision and high-performance machining, ISO standard inserts could not machine quality bores, rough and semifinish cast iron or steel in one last finishing pass, or finish-machine with precision. Modifications to an ISO standard tool provide a more precise insert pocket and permit adjustability, as shown in Figure 5-27. The insert pockets have a 0.39–0.79 mil (10–20 μm) angular accuracy. The adjustment device consists of a precision, adjustable clamping plate and a differential screw. When moving the clamp plate, the angle surface of the blade is pressed outward using the slight air gap between the clamp screw and bore in the indexable blade and the elasticity of the clamp screw. Adjustments of up to 0.004 in. (0.10 mm) can be achieved. This allows for defined, accurate ISO-insert setting with tools that feature multiple arrangement of inserts, both radially and axially.

In machining, tool adjustments are often necessary, sometimes at the expense of productivity. The objective is to find the right balance between the two.

# STRINGENT FINISH REQUIREMENTS

In manufacturing there is a growing need for machining tighter toleranced parts. The average manufactured product of just 15 years ago would fall short of today's requirements. The quality of a finished, manufactured product is the reflection of its workmanship: the way the individual parts are finished, fit together as a whole, and function. A structured and methodical engineering approach is used to express physical sizes, characteristics, and features of parts with pre-described tolerances. Tight tolerances and excellent surface finishes are pivotal manufacturing objectives that should be considered.

To meet specified part finishes, every manufacturing company must:

- explore the simplicity of functional and physical part characteristics;
- select part material with low-cost production in mind;
- seek the most productive and economical manufacturing process; and
- specify geometric finishes commensurate the type of part, the production equipment, and systems available.

## Tight Tolerances and Their Effect

In the wake of mass-produced products and assembly-line set-ups, it became clear that parts and components had to be dimensionally defined and toleranced to make them fit and in random order. The aspect of allocating certain male and female parts and their interchangeability added to the concept of tolerancing. Debugging parts and production methods based on trial and error was no longer affordable. To make automatic assembly work, the parts had to be produced with much closer tolerances. Controlled, closer tolerancing made products more reliable. The pressure was strong on keeping manufacturing costs down and on being competitive. Doing it right the first time had become the only sensible approach. Today, other equally important demands have to be met:

- improved safety,
- energy savings,

- weight and noise reduction,
- emission control, and
- extended service life.

Design engineering, as a result, has to specify even tighter tolerances for practically all geometric part characteristics:

- flatness,
- roundness,
- straightness,
- angularity,
- perpendicularity,
- parallelism,
- concentricity,
- symmetry,
- surface and line profile,
- runout, and
- position.

In addition, surface-finish requirements increase. For example, the tighter the tolerances of two mating parts, the smoother the surface required. In fact, smooth, even surfaces are essential in the areas of:

- material fatigue strength,
- corrosion resistance,
- sealing performance,
- friction,
- lubrication, and
- force distribution.

Tolerances and surface finishes usually complement each other. While the scope of controlled tolerancing is widely known, there seems to be less certainty as to some areas of surface technology.

## Surface Texture and Integrity

Surface texture describes the quality of a workpiece surface through three different characteristics: roughness, waviness, and lay. The workpiece material, machine tool, cutting-tool system, machining data, and machining periphery all affect the surface. Texture roughness is universally used to indicate a certain finish requirement, as shown in Figure 5-28. It is defined through different parameters and equations and given in numerical values. The approximate correlation is:

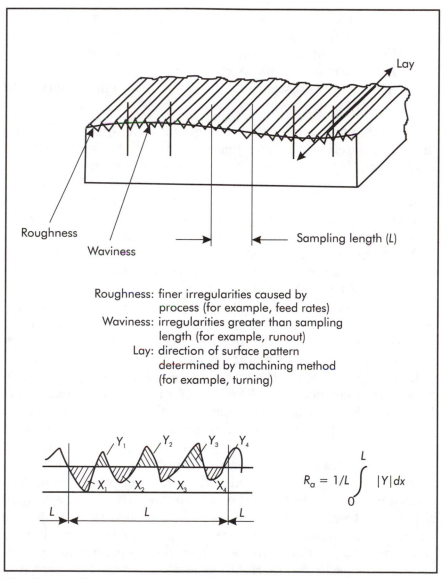

Figure 5-28. Surface texture and integrity.

$$1\ R_a \simeq 40\ RMS;\ R_t \simeq (6\ \text{to}\ 7)\ R_a \qquad\qquad (5\text{-}1)$$

$$R_z \simeq 0.85\ R_t$$

where:

$R_a$ = arithmetic average height from the mean (most commonly used)

$RMS$ = geometric average height from the mean

$R_t$ = maximum peak to valley height

$R_z$ = average distance between the five highest peaks and five lowest valleys

For most applications, and this includes high-precision manufacturing, the criterion of roughness as a measure of surface quality is sufficient. However, in cases where specific surface patterns are required, the surface texture needs to be defined, specified, and controlled. An example would be a bearing surface specified with a certain roughness profile (a certain peak and valley distribution) for better oil retention, rather than low, average roughness. Different machining methods yield different surface textures. In specific cases, surface textures have to be engineered and machined to provide functional surface characteristics.

Every material removal process has its own effect on the workpiece. While surface texture is the criterion for workpiece finish, surface integrity checks the interior effects of the machining processes on the workpiece material.

Critically stressed parts should have surface integrity specifications that consider the alteration of surface layers (plastic deformation) by the machining process and the effects this has below the geometrical surface of the workpiece. The machining process can result in micro-cracks, recrystallization, residual stresses, hardness alterations, and non-uniformity of the material. The number of workpiece material and machining process combinations is countless. However, the two surface-integrity effects with the most direct bearing on design and application are fatigue strength and stress corrosion cracking propensity. Both, in many cases, need to be empirically determined for a particular application before specifying the right material, machining process, and finishes.

# COST AND QUALITY

Over-specifying at the design stage and using the most intricate, expensive machining methods do not necessarily improve a product's quality. In fact, the design has to be function- and manufacturing-productivity oriented.

Manufacturing cost increases with the number of machining passes, and the specification of tighter tolerances, and better part finishes. The other side of the spectrum is that manufacturing without the proper processes, methods, and finishes can be even costlier, due to premature part failure or shortened product life. Over-specifying and under-specifying are both unacceptable.

Are there any cost guidelines engineers can go by? The truth is, many attempts have been made to empirically find cost guidelines. The upshot is tolerance increments plotted over relative production cost describe a reversed parabolic curve. This means production costs increase geometrically for uniform tightening of tolerances. However, relative cost is just that, relative. The same holds true for the relationship between machining cost and surface roughness. Tight tolerances and good surface finishes can be achieved with advanced cutting-tool systems at no extra production cost.

The key is for design and manufacturing engineering to avoid extremes. Narrow specifications should be applied selectively only where they are needed, rather than for the entire workpiece. A dark shadow on an otherwise smooth, per specified surface finish, does not necessarily call for rework because of a nonfunctional, visual flaw. As to surface texture, the criteria for quality is not always the smoothest because machined surfaces are complex and parts with different functions sometimes require different texture profiles. The right distribution of peaks and valleys has to be specified, especially when friction, oil retention, or adhesions are required. Coarser finishes are sometimes more desirable.

Surface-integrity specifications should be specified only on the critical and highly stressed zones of the component part. Whenever high tensile strength and high stress resistance to extreme and alternating conditions are relevant factors, the machining processes become more involved, and cost and productivity take a second seat to safety. It is not only the machining method that causes the costs to increase. Raw, unmachined material, post-machining processes, as well as an overly cautious machining approach (conservative machining data) add to production costs.

## Achievable Parameters

Industry-wide and the world over, precision-machined parts are dimensioned, produced, and inspected for quality. From design to manufacturing there needs to be a real understanding of what it means to manufacture to within micrometers of specification (see Figure 5-29). Before specifying tight tolerances for any part, the following criteria should be considered:

- function of the part;
- manufacturing capabilities;
- industrial quality standard;
- cost;
- production volume; and
- measuring equipment.

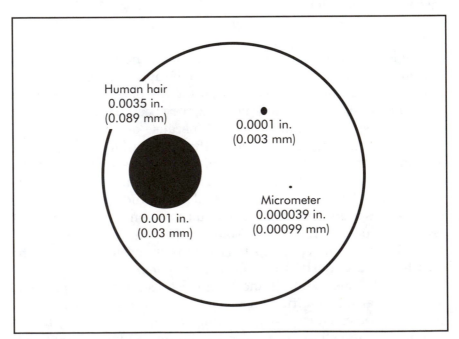

Figure 5-29. The micrometer compared to other dimensions.

Functional, not visual aspects dictate the narrowness of the finishing parameters. Manufacturing capability is determined by the whole machining and manufacturing system. Often overlooked is the fact that in many cases the tolerance sum of the individual components of the machining system is greater than the specified part tolerance.

Capabilities have to be identified to specify feasible finishing parameters. Technological progress in the areas of metallurgy, machine tools, and cutting tools has substantially narrowed down international industry tolerance standards for respective machining operations and component parts. For example, IT5, IT6, and IT7 are current international tolerance bands for finish-metal-cutting processes.

Assuming specified close tolerances are constant and production volumes variable, the manufacturing approach certainly would be different. Holding close tolerances in a high-volume production environment calls for special machining processes and methods. Here quality and consistency are more difficult to achieve.

How about specifying finishes that cannot even be verified? Manufacturing of close-toleranced parts is an expensive proposition. It is essential to have measuring equipment available that is capable of measuring up to the quality of finishes produced. Verification at the end of the production process ensures the all-important feedback of how manufacturing is doing and whether the part finishes demanded can be met.

## TOTAL QUALITY MANAGEMENT

The purpose of inspection at large is acceptance or disposition of product based on its quality. There are three types of decisions to be made:

1. conformance decision: judging whether the product conforms to specification;
2. fitness-for-use decision: deciding whether nonconforming product is fit for use; and
3. communication decision: deciding what to communicate to insiders and outsiders.

The principles of the "first good part" and "zero defects" must be pursued. This means to strive for almost perfect results, eliminating inconsistencies, rework, and foregoing scrap. To accomplish this, total quality management (TQM) must be embraced by all departments of the enterprise.

Statistical process control (SPC) methods are part of TQM. They were developed in response to the realization that variations in processes and products exist. Indirect variations in production processes can cause the same end products to differ from one another. Recognizing certain variability as inherent helps to decide upon the disposition of the product.

## Statistical Measures

Manufacturing has to be done with predictability and consistency to maintain quality levels. A high quality level is possible with well-designed and well-executed systems and processes. Quality performance has to be measured and compared to standards. Deviations need to be acted upon. This is called *quality control*. Six-sigma techniques and process capability analysis numerically describe the quality of product and process.

Six sigma, process capability, and true position are statistical measures to evaluate the outcome of a manufacturing process, compare it to a standard and/or a predetermined objective, and reveal possible causes of nonconformance.

### Six Sigma

The mathematician Gauss proved the existence of natural distribution. Every condition or event in life follows a certain pattern of variation, known as the bell curve, as shown in Figure 5-30, which also can be applied to manufacturing processes.

When a product or process exhibits small variations, the sigma level is high and the likelihood of defect low. Six sigma is accepted as the highest level of manufacturing quality. It covers 99.7% of the area of normal distribution. When a product is "six sigma" it exhibits 3.4 defects per million, taking the typical variations of manufacturing into account. Six sigma, in other words, is a predetermined quality standard by which manufacturing has to produce its parts. As a control device, it is the area of the bell curve

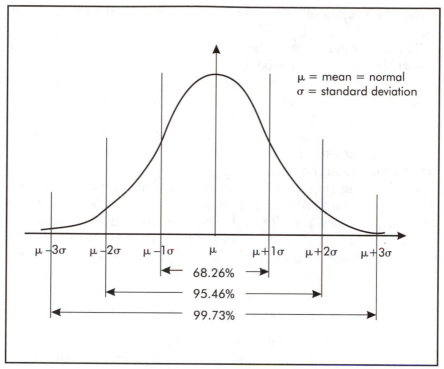

$\mu$ = mean = normal
$\sigma$ = standard deviation

$\mu - 3\sigma$    $\mu - 2\sigma$    $\mu - 1\sigma$    $\mu$    $\mu + 1\sigma$    $\mu + 2\sigma$    $\mu + 3\sigma$

68.26%
95.46%
99.73%

*Figure 5-30. Bell curve.*

wherein sample parts, taken at random, have to fall at the end of the machining process.

Quality control, when sampling parts after manufacturing, typically does so by forming a histogram, a vertical bar chart of frequency distribution. The histogram is like the distribution curve. Highlighting the center and variations of data is popular because of its simplicity in use and value of interpretation. The shape of the histogram, plotted within the specification limits, can show the process condition.

## Process Capability ($C_{pk}$)

Process capability measures the uniformity of machining and manufacturing processes by evaluating machine tools and cutting tools through the results achieved. Process capability analysis identifies variations of the process from the nominal. To illustrate, graphs of two example processes are shown in Figure 5-31.

The target diameter of the workpiece is 5 in. (127 mm) with an allowable tolerance of ±0.002 in. (±0.05 mm). That is, a maximum diameter of 5.002 in. (127.05 mm) and a minimum diameter of 4.998 in. (126.95 mm). A sample of 50 parts taken off the machine, their diameters measured and then plotted down, yields a histogram, which reveals that all parts fall within a range of +3 sigma standard deviation. In both cases, 99.97% of all parts fall within the six-sigma limit.

$C_{pk}$, by definition, is the measurement of how well manufacturing and machining centers are performing. As can be seen in Figure 5-31, Process 1 is right on target. Process 2, by contrast, is moved off center to the lower end of the specified tolerance. This offset from center is precisely what determines process capability. It is defined mathematically as follows:

$C_{pk}$ = allowable tolerance range
$2x$ = farthest away from the mean

Figure 5-31 shows that Process 1 yields a $C_{pk}$ of 2.0, and Process 2 a $C_{pk}$ of 1.33, which has been accepted by most industrial manufacturers as a desirable and acceptable objective, while trying to work toward a higher quotient. If a machining process cannot meet the expected capability, machine tools might have to be overhauled, or different ones specified, or other cutting-tool systems used (Juran and Gryna 1989).

## True Position

True positioning, as a statistical tool in quality control, is applied where tight tolerance paired with function and interchangeability of mating parts are crucial. The center of the bore, by definition, is its true position, which is determined by distance specifications from another location on the part, for example, manufacturing bores. The example in Figure 5-32 shows high accuracy within tolerances. A comparative measurement of several sample parts can find the actual location of the bore by plotting the deviations in the $X$ and $Y$ axes. The average total deviation stipulates the within or out-of-tolerance condition.

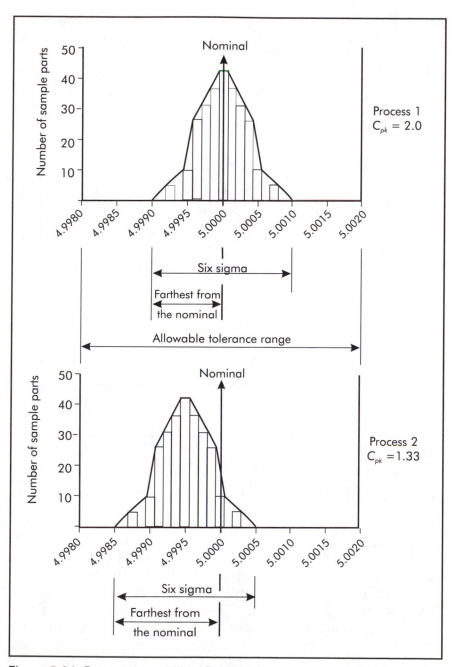

Figure 5-31. Process capability ($C_{pk}$).

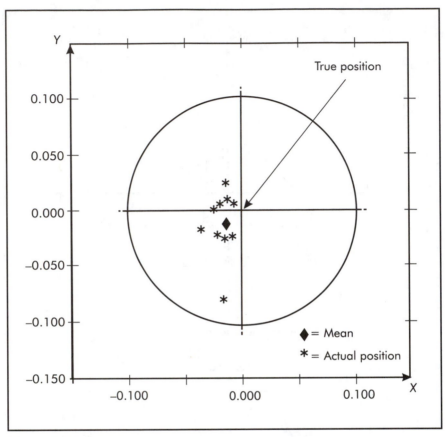

*Figure 5-32. True position.*

## INTELLIGENT TOOLING

Designing features into cutting tools that monitor the machining process and communicate the status has been a pursuit by some manufacturing companies. In-process sensors are used to automatically control machining processes. Acoustic, force, and vibration sensors detect critical degrees of chip formation, surface deterioration, excessive tool wear, variation in machining data, etc. Tooling manufacturers can choose from many types of sensors. The choice depends on the application and the accuracy and difficulty of the monitoring process. Using sensor, transmitter, and electronic technology with subprograms residing in the machine control system creates intelligent, high-precision, highly productive cutting tools. A schematic of the closed-loop electronics for such a tool is shown in Figure 5-33.

With the control and electronic model, energy and data transmission do not take place by contacts but by induction. The built-in servomotor actuates the motion of the tool and there is a four-fold oversampling at 800 kHz for fast and trouble-free operation, as shown in Figure 5-34. The result for this tool design is an additional machine axis that can perform rather complex contours.

An intelligent tool of a different kind is designed for a machining center with integrated compensation and gaging, as shown in Figure 5-35. The machining is performed with three inserts spaced 120° apart. Two inserts are fixed and semi-finish the bore. The third insert is set to the finish dimension. The blades are arranged for balanced stock distribution in such a way that the tool can be

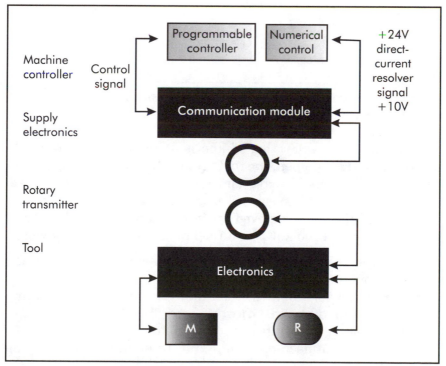

*Figure 5-33. Closed-loop control electronics.*

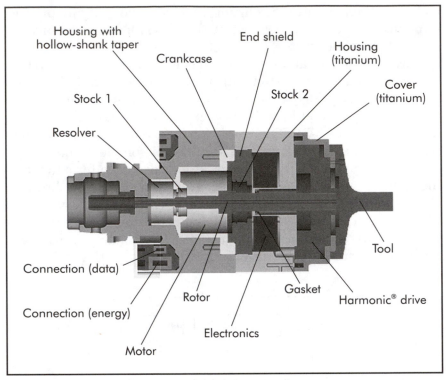

*Figure 5-34. Electronically activated feed-out system.*

quickly retracted when the spindle is stopped with an offset to the rotational axis. Then, the tool probes the bore size through an integrated air gage. If corrections to the finish diameter are necessary, the integrated mechanism is activated to adjust the setting diameter. This is achieved through a combined activation of rpm increase and coolant pressure. The advantages are obvious:

- reduced cost due to elimination of expensive compensation components built into the machine;
- integrated size compensation due to insert wear with accuracy within 39.4 μin. (1 μm);
- Automatic gaging integrated in the tool;
- reduced non-machining time due to elimination of gaging step; and
- Semi-finishes and finish operations in one tool.

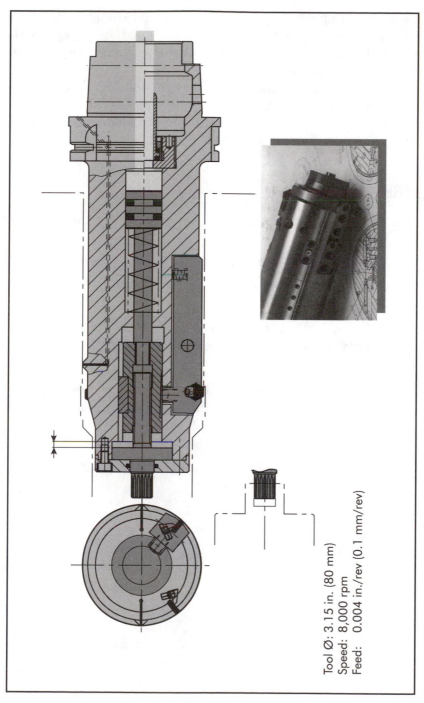

Tool Ø: 3.15 in. (80 mm)
Speed: 8,000 rpm
Feed: 0.004 in./rev (0.1 mm/rev)

*Figure 5-35. Actuating tool. (Courtesy MAPAL, Inc.)*

There is an optional design with adjustment through a hex nut, which is mounted inside the working area. The tool engages with the nut and performs the necessary number of turns to adjust the radial setting dimension.

The future will see breakthrough designs for so-called "intelligent tooling." Such tooling will communicate with the machine control system and adjust itself in automatic mode to variations during critical machining processes.

## REFERENCES

*Cutting Material and Tools*. 1989. Dusseldorf, Germany: VDI-Verlag GmbH.

Dornfeld, David A. 2001. *Research Reports 2000-2001*. Berkeley, CA: University of California-Berkeley, Laboratory for Manufacturing Automation.

Hwang, Edward I. 2001. *Research Reports 2000-2001*. Berkeley, CA: University of California-Berkeley, Laboratory for Manufacturing Automation.

Juran, J.M., and Gryna, Frank M., Jr. 1989. *Quality Planning and Analysis*. New York: McGraw-Hill Publishing Company.

# Precision Tooling Interface

# 6

Every element of the machining process must be equal to each of the other components' level of accuracy, quality, and technology. The goal is to integrate the machine tool, cutting tool, peripheral machining elements (fixturing, coolant, etc.), and tooling interface with one another. The machining process as a whole cannot be of high precision/high performance, if any one element is uncoordinated (see Figure 6-1).

For too long, the tooling interface, the connection of cutting tool and machine tool, has been overlooked, underestimated, and ignored. There are several tooling interfaces that will eventually

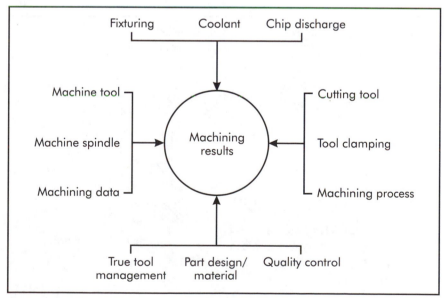

Figure 6-1. Major influences on machining results.

become the standards for high-performance machining. These interfaces use a system known as the hollow-shank (HSK) taper. The interfaces offer the superior characteristics of:

- high torque transmission,
- high repeatability accuracy,
- excellent adaptation to high cutting speed,
- uniformity,
- minimal runout,
- high clamping force,
- reliability and safety,
- static and dynamic stiffness,
- ease of handling, and
- excellent adaptability.

## CONNECTION AND INTERFACE

High-speed precision cutting can be a problem with the existing technology. HSK tooling was developed to eliminate the inherent design disadvantages of the CAT (Caterpillar) V-flange (see Figure 6-2) at extremely high spindle speeds.

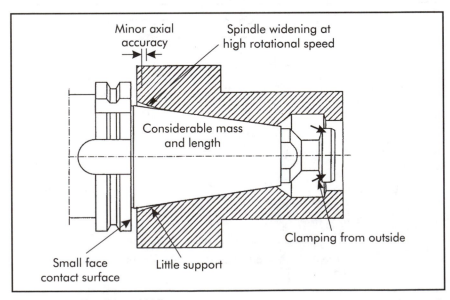

Figure 6-2. Traditional V-flange.

Seldom has such a small cutting-tool innovation had such a great impact on machining as the hollow-shank taper. In fact, it has revolutionized not only the design of precision cutting tools, but also the approach to high-cutting-speed machining and tooling interface and clamping systems, as shown in Figure 6-3.

The HSK taper is best for meeting the demands for a precision tool connection and interface. Due to its exact taper and relatively large face-contact area, the HSK taper exhibits high repeatability accuracy and practically no angular runout. This is especially important for flexible, computer numerical control (CNC) machining where tools are moved in and out of the machine spindle repeatedly.

Extensive research and test runs have determined that the HSK taper has a stiffness seven times higher than the steep-shank (SK) or CAT taper. The torsional load capacity with HSK compared to

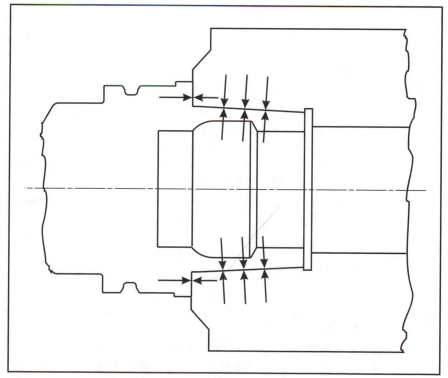

*Figure 6-3. HSK in clamping position.*

the conventional taper shank is also much higher, as shown in Figures 6-4 and 6-5.

One of the most important factors for high-performance machining is rotational speed. In this case, HSK proves to be the perfect design. As the spindle speed increases, the taper of the HSK expands against the mating contour of the clamping chuck. The centrifugal force on the clamping segment increases the clamping force and creates a very tight fit. The taper shank and the machine spindle opening have specified tolerances for radial over-dimension. This causes pressure clamping in addition to form clamping, which gives a high transmission of torque. The maximum allowable rotational speeds for HSK have been empirically determined to be 40,000–45,000 rpm.

HSK shanks 32, 40, 63, and 100 are standard designated sizes, covering a full range of capabilities for practically all machining operations. Their design invites modularity and, since HSK has become an ISO standard, they are generic and can be manufactured by any cutting-tool manufacturer. However, their precise manufacture must meet the high standards of the HSK connection and interface.

The benefits of the HSK taper over conventional tooling include:

- It is highly suited for the varying machining conditions—from heavy stock removal to high-cutting-speed machining.
- It provides high changing and repeatability accuracy.
- It can be used for all popular machining operations such as turning, boring, drilling, reaming, threading, etc.
- It can be equally well adapted to manual and automatic clamping.
- Its innovative design has filled the technology gap between the high-performance machine tool and high-performance cutting tool (Lembke 1997).

## TOOL CLAMPING

There are three tool-clamping systems in use in industry: mechanical, hydraulic, and thermal. The merits of these different systems are shown in Table 6-1. The user has to decide which one

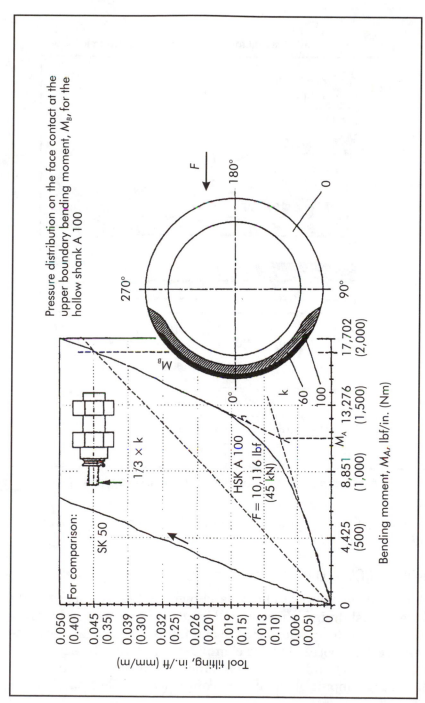

*Figure 6-4. HSK maximum bending moment.*

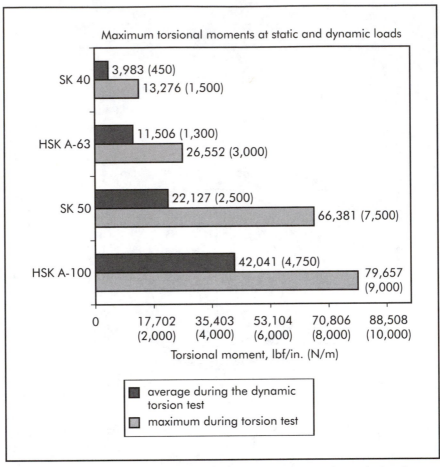

*Figure 6-5. Maximum torsional moments at static and dynamic loads.*

is best suited to the application, depending on the respective machining conditions and variables.

## Mechanical Clamping

A typical mechanical chuck is shown in Figure 6-6 with its simple, robust design, short overall length, and short flow of force, which offer uniform and accurate clamping. The chuck's key elements are incorporated in the clamping cylinders. The extraction (clamped mode) and contraction (unclamped mode) of the four, half-round clamping elements, are controlled via a left-hand/right-

#### Table 6-1. Comparison of different clamping systems

**Mechanical Clamping (HSK based)**

No presetting necessary
High torque capacity
Best total indicator runout; no runout stack up between HSK and cylinder shank
Most rigid design
Fast tool change, no special/expensive equipment needed

**Hydraulic Clamping**

Fast tool change, no special/expensive equipment needed
Can be used with sleeves
Reduced inventory
Length adjustable
Improved tool life through dampening effect

**Thermal Clamping**

Up to four times holding power over hydraulic chuck
No hardware except for length-adjusting screw
No maintenance
Length adjustable
Inexpensive holder

hand screw. An extraction pin disengages the tool from the chuck securely and quickly during unclamping. Large, self-contained coolant ports provide central coolant supply. A bayonet lock holds the clamping cylinder firmly in position, as shown in Figure 6-7.

A short flow of force guarantees a very rigid connection with extremely high clamping pressure up to 900 lbf (4,000 N), and accuracy between 0.12–0.16 mils (3.0–4.0 μm). The mechanical-chuck is most popular as a flange-mounted adapter or an in-between adapter as part of a tool assembly or in conjunction with the CAT-V-taper shank for machining centers (see Figure 6-8).

The ease of handling of repeatable mechanical chucks mated to HSK tapers is testimony of the simplicity and high adaptability of HSK. The most important and basic feature of the HSK tapers and HSK-based chucks is their precision of manufacture. Extremely tight tolerances and accurate measuring equipment must be applied to secure their proper function. The unparalleled success of HSK in high-performance-production machining has

*Figure 6-6. Mechanical clamping chuck.* (Courtesy MAPAL, Inc.)

enticed cutting-tool manufacturers to develop small-diameter tools in the 0.20–0.59 in. (5.1–15.0 mm) range with a shank based on the merits of HSK tapers. The features of such tooling include: even axial drawing-in of the tool, extremely precise taper, large contact face, and precise mating of the clamping device.

Built-in HSK clamping performs especially well in critical applications dominated by extremely high speeds, high metal-removal rates, high-tolerance operations, and frequent and fast tool changes.

Figure 6-9 shows a symmetric clamping prong with axial motion of the clamping elements controlled via two clamping tapers. The force, produced by the spring assembly (or other energy carrier, such as hydraulic pressure), is transferred to the clamping taper through a drawbar. The clamping taper transfers the force to the collet pieces at the circumference at which the tool is clamped. Because of the increased force of this system, it is possible to decrease the spring-disc assembly by a factor of three to

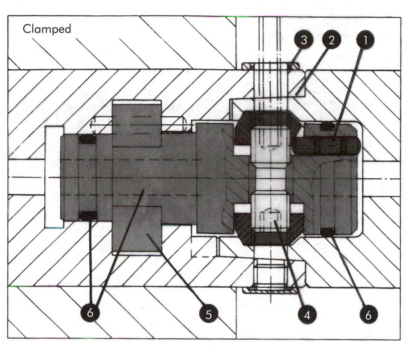

Clamped

1. Automatic ejection of the shank when the chuck is opened. This is particularly important for the self-locking HSK connection.

2. Increased pretensioning of taper allowing maximum accuracy

3. Automatic sealing of the clamping hole

4. Can be opened even where a hexagonal edge of the clamping screw has been damaged

5. Easy-to-change clamping cartridges with bayonet fitting

6. Central coolant supply, fully sealed

*Figure 6-7. Principle of mechanical chuck.* (Courtesy MAPAL, Inc.)

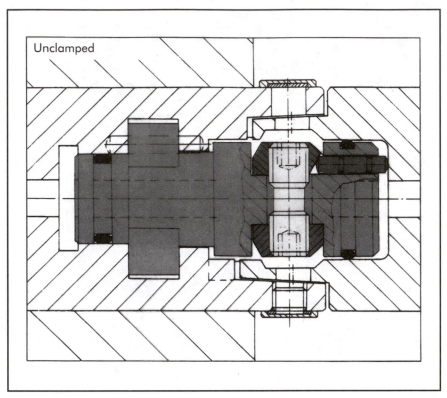

*Figure 6-7. continued.*

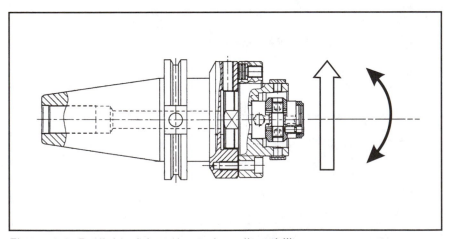

*Figure 6-8. Radial, axial, and angular adjustability.*

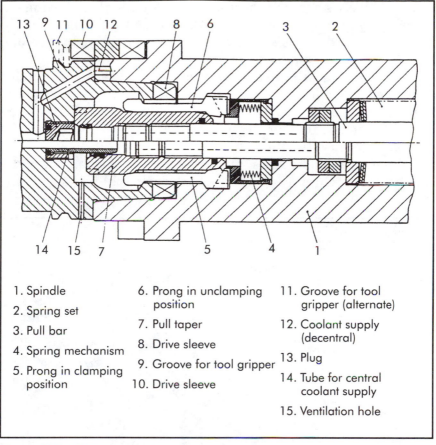

*Figure 6-9. HSK spindle design.*

1. Spindle
2. Spring set
3. Pull bar
4. Spring mechanism
5. Prong in clamping position
6. Prong in unclamping position
7. Pull taper
8. Drive sleeve
9. Groove for tool gripper
10. Drive sleeve
11. Groove for tool gripper (alternate)
12. Coolant supply (decentral)
13. Plug
14. Tube for central coolant supply
15. Ventilation hole

four when compared to conventional steep-taper spindles. This is made possible through a preloaded face-contact surface. This adds to the rigidity of the spindle and allows smaller spindle designs. For automatic HSK-based clamping, the spring-disc design is as good as the hydraulic configuration.

## Hydraulic Clamping

Hydraulic tool clamping has become one of the most popular methods for high-performance machining, especially for drilling, boring, and reaming operations. Despite intricate inner contours required to expand and contract hydraulics, the clamps are built with relatively small and short overall dimensions, achieving extremely high clamping pressures of up to 3,000 psi (211 kgf/cm$^2$). They accommodate tools with round shanks, holding them along the clamp's axial centerline, as shown in Figure 6-10.

Hydraulic chucks perfectly transmit axial pushes and torques generated when drilling. For extensive side loads that occur during milling operations, the hydraulic chambers can be too flexible. But, it is this feature that makes the hydraulic chucks desirable because the fluid chambers have a natural dampening effect. This minimizes vibration, which facilitates balancing of the tool assembly. Often overlooked are the minimal but important

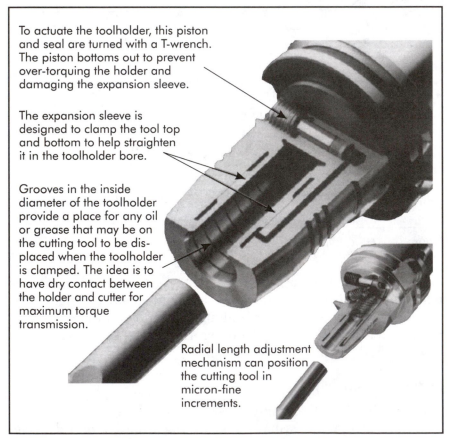

To actuate the toolholder, this piston and seal are turned with a T-wrench. The piston bottoms out to prevent over-torquing the holder and damaging the expansion sleeve.

The expansion sleeve is designed to clamp the tool top and bottom to help straighten it in the toolholder bore.

Grooves in the inside diameter of the toolholder provide a place for any oil or grease that may be on the cutting tool to be displaced when the toolholder is clamped. The idea is to have dry contact between the holder and cutter for maximum torque transmission.

Radial length adjustment mechanism can position the cutting tool in micron-fine increments.

*Figure 6-10. Hydraulic clamping.* (Courtesy Modern Machine Shop)

maintenance needs for hydraulic chucks. For example, screws have to be kept tight and seals properly maintained. Only then can the holders' high concentricity and uniform clamping pressure be maintained. Designing a hydraulic chuck with an HSK interface makes the holder very stable axially and radially, and ensures high repeatability accuracy for frequent tool changing.

## Thermal Clamping

Thermal clamping, also known as shrink-fit clamping, is accomplished by induction heating the toolholder and then shrink-fitting the cutting tool into the holder while it cools down. With this method, the toolholder and tool become as close an integral part of each other as possible, with a 360° grip over the entire shank length.

Applied with the interference-fit principle, whereby the holder's inner diameter is deliberately undersized, the holder is thoroughly heated, usually by induction heating and air. When a temperature of 600–800° F (316–427° C) is reached, the inside diameter of the toolholder expands sufficiently to allow insertion of the tool within 7–10 seconds. When cooling down, the toolholder grips the cutting tool firmly and uniformly with a high force of up to 10,000 lbf (44 kN). The concentricity of the cutting tool, set in the shrunk condition, is about 0.08–0.20 mils (2.0–5.0 μm). Shrink-fitting results in high rigidity due to the strong holding force. Because of the cutting tool's concentricity and balance, chip loads during machining are evenly distributed on the cutting edge, extending tool life. The simplicity of the cutting tools being mated to the holders (including replacing tools) is one of the advantages of the system. However, with the variety and complexity of cutting tools increasing, quick and accurate presetting may be required. Some systems combine the heat-induction device with tool presetting in one operation. They have become popular in higher volume, precision machining involving different cutting-tool materials, step tools, and counterbore tooling.

Figure 6-11 shows a device that shrinks, presets, and inspects the tool. First, the shrink-fit holder is placed into the heating device and raised to the desired temperature. Then, the cutting tool is put into the holder and brought into position for the tool length

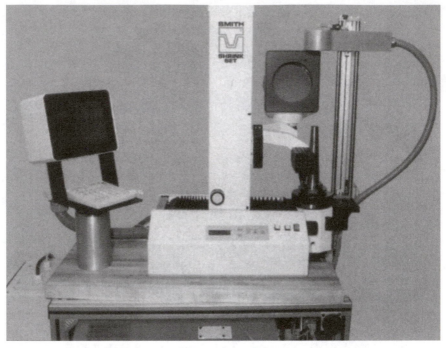

*Figure 6-11. Shrink-fit system.* (Courtesy T. M. Smith)

required. Finally, it is compared with the template of the unit's optical system.

## BALANCING

Balance is always a concern with toolholders, although the balance of the entire tooling assembly is really the yardstick to go by. A well-balanced toolholder mated to an unbalanced cutting tool and/or adapter creates an assembly with an out-of-balance condition. Balanced toolholders and better cutting-tool assemblies eliminate vibration and tool chatter, and ensure uniform chip loads. The results are increased tool life, decreased machine-spindle load, and consistent quality of part-surface finishes, as well as faster material removal rates. A perfectly round, rotationally symmetric design with a superfinish would show no or minor dynamic imbalance. Invariably, because of their inherent designs, cutting tools and toolholders always register certain imbalances. The question is what is acceptable for workpiece finish, tool life, spindle life, and safety.

The relationship between centrifugal force and speed is squared. That is, doubling the speed quadruples the centrifugal force. The American National Standard Institute (ANSI) defines the *acceptable imbalance* of the rotating body relative to the maximum service speed. ANSI assigns balance quality grades to related groups of rotating bodies, designated as G-numbers, as shown in Figure 6-12.

*G-numbers* are a measure of imbalance. The numbers are obtained by multiplying the distance in millimeters from the tool's geometric centerline by its actual rotating centerline. Then, multiply the product times the rotational speed, which equals rpm. Quality grades of between 2.5 and 6.3 are what manufacturers of machine spindles, toolholders, and cutting tools should deliver. It is very important that all elements involved in the machining process show a uniform balance grade. If this is followed early in the process, huge problems will be avoided later.

# RUNOUT

The system comprising of machine tool, toolholder, tool clamping, and cutting tool invariably and inherently calls out a certain runout, which is measured from the spindle face to the tip of the cutting-tool insert. The main contributing components are the:

- machine tool, which includes the spindle bearing, spindle diameter, and manufactured accuracy;
- toolholder/tool clamping, which includes radial stability, repeatability, stiffness, clamping force, and manufactured accuracy; and
- tool body, which includes tool balance, material, length-to-diameter ratio, thrust-face diameter, and manufactured accuracy.

Modern, state-of-the-art machine tools, toolholders, tool-clamping systems, and cutting tools offer precision of lesser or higher degrees. In high-precision manufacturing the runout of the whole

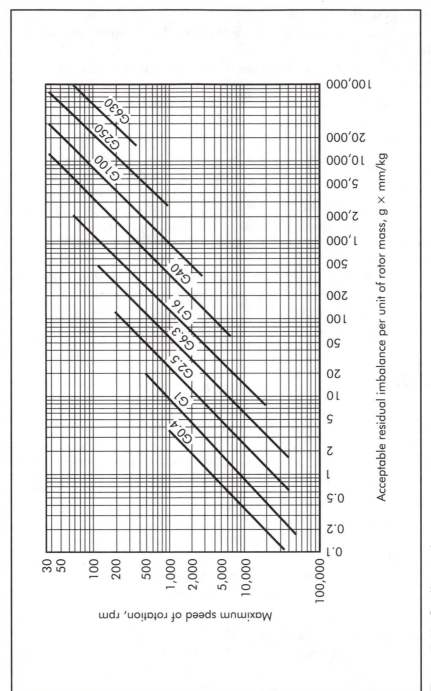

*Figure 6-12. Quality grades.*

machining system has to be kept to a minimum, striving for zero (0). The individual errors are a deviation of true radial and angular position relative to the theoretical centerline.

Mounted in-between the toolholder and cutting tool, the adapter features two separate bolt circles, each with four bolt holes. When manually rotating the cutting-tool system installed in the spindle, a dial indicator, touching the tip of the insert indicates the deviation from the theoretical centerline. With four bolts on the periphery, the tool system can be reduced to zero, eliminating both radial and angular runout (see Figure 6-13). Provisions for angular correction have to be made, particularly for longer tools exceeding length-to-diameter ratios of 8:1. Once the whole system is clocked-in and showing zero runout, machining can begin.

## REFERENCES

Bakerjian, Ramon and Philip Mitchell, eds. 1993. *Tool and Manufacturing Engineers Handbook*, Fourth Edition. Volume 7, *Continuous Improvement*. Dearborn, MI: Society of Manufacturing Engineers.

Lembke, Dietrich. 1997. *Information Management*. Frankfort, Germany: CIM GmbH.

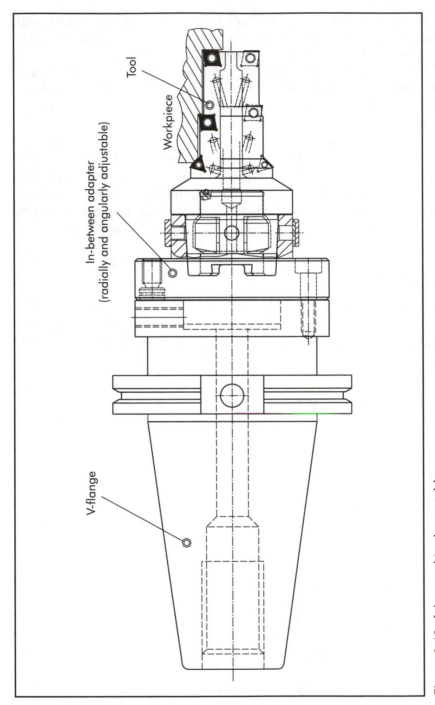

Tool

Workpiece

In-between adapter
(radially and angularly adjustable)

V-flange

*Figure 6-13. Advanced tool assembly.*

# Manufacturing Enablers

# 7

Some manufacturing enterprises realize that the supply of products and services by a third party can have an enormous impact on the bottom line. If done right, it can be very beneficial. Unfortunately, many manufacturing companies do not know how to take advantage of their relationships with suppliers. In not doing so, they are at risk of compromising product innovation and technology, and potentially losing market share. Some do not realize the potential value of outsourcing and full-service supply, ignoring a powerful tool of competitiveness (Chopra and Meindl 2000).

## THIRD-PARTY SUPPLY

Outsourcing, one of the more recent corporate strategies, carries a certain mystique with it. Experts have often heralded it as a new panacea discovered overnight through someone's intuition. However, its working principle was around in the private sector even before mankind's first step toward modern society. A typical example of personal outsourcing is having someone else do certain chores such as child care, house cleaning, maintenance of home and property, etc. This allows the homeowner to pursue other activities to fulfill the family's wants and needs, such as earning money, engaging in sports activities, health care, and so on. Industrial outsourcing grew out of this; it is a symbiotic relationship borne out of necessity.

Outsourcing of parts or logistics of tooling for production purposes has occurred for essentially these reasons:

- The original equipment manufacturer (OEM) wants to concentrate on core business (essential competency activities).

- The complexity of processes and products is increasing.
- The experts or specialists are better at it.
- The OEM's labor costs are too high.
- The OEM wants to take advantage of just-in-time (JIT) delivery.
- There is too much money tied up in inventory.
- The principle of outsourcing creates a business partnership, built on trust, reliability, and mutual benefits.

## CHANGE OF BUSINESS PARADIGMS

Ever since the late 1940s, political and economic models have rapidly changed. The way business was perceived and conducted had an effect on manufacturing. Interestingly, most factors influencing corporate strategies now are diametrically opposed to what they once were (see Figure 7-1).

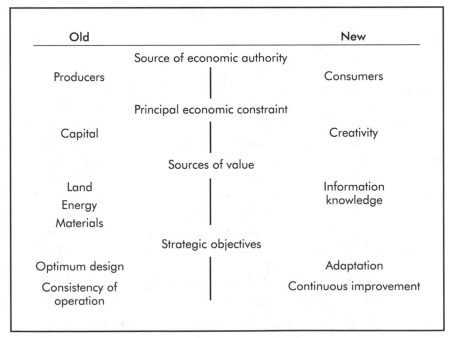

| Old | | New |
|---|---|---|
| **Source of economic authority** | | |
| Producers | | Consumers |
| **Principal economic constraint** | | |
| Capital | | Creativity |
| **Sources of value** | | |
| Land | | Information |
| Energy | | knowledge |
| Materials | | |
| **Strategic objectives** | | |
| Optimum design | | Adaptation |
| Consistency of operation | | Continuous improvement |

*Figure 7-1. Business paradigms.*

The industrial revolution and subsequent mass production allowed producers to dictate product design, options, delivery, and price to buyers for about a century. However, consumers are now telling the producers what to design, the options, and terms of delivery. The source of economic authority has clearly changed from the producer to the consumer.

The principle economic constraint used to be the availability of capital. Today, it appears that innovation and technology determine the well-being of economies. Knowledge and information have begun to add greater value to micro- and macro-economics than land, energy and materials—the traditional sources of value. Corporate strategies of optimum, uniform designs and operational consistency have given way to agility, the timely adaptation to changes, and the perpetual quest for process and product improvements. These shifts and changes have led to seemingly insurmountable pressure on the OEM. With customers' demands increasing, more innovations, increased levels of knowledge, and the need for more rapid adjustments, there is a natural need for assistance and relief on the production floor.

## Making the OEM/Supplier Relationship Work

When OEMs decide to outsource products or services and the supplier agrees to deliver, the relationship must be based on certain building blocks.

First, there must be trust. The supplier must know that the manufacturer is sincere and committed to the program, preferably through contractual agreements. The supplier, in turn, must be knowledgeable enough to provide what the manufacturer expects. Better yet, the supplier should provide the product or service that represents the best that can be provided at the time of purchase (cost, quality, time).

Second, both parties need ongoing communication in place to provide up-to-date information to each other. Continuous improvements of manufacturing processes and products are important aspects of the OEM/supplier arrangement. This assures the OEM of further development and possible technical advancement, and allows the supplier to provide innovative and progressive new ideas.

The best vehicle for advancement, teamwork, suggestions, and the creation of new ideas is concurrent engineering and its extension, concurrent manufacturing. It can be the forum for brainstorming and benchmarking. Interdisciplinary teams from both sides systematically strive for continuous improvement. Concurrent engineering's importance in this relationship can not be over emphasized. It eventually opens the door to concurrent manufacturing, a prerequisite for success involving sister companies, satellite plants, and simultaneous offshore manufacturing.

Concurrent engineering leads to the third basic building block, partnering. For the supplier, this means becoming an extended part of the OEM's enterprise. The mutual strive for consistent, top-quality product, more efficiency, and higher productivity will potentially make the supplier world competitive and the OEM a best-in-class manufacturer, as shown in Figure 7-2. Working closely together and regarding their relationship as a foundation for continued growth, both must be continuously involved in the process.

## Selecting the Right Supplier

When selecting a supplier, an OEM usually requires that:

- The supplier knows the management philosophy of the OEM and continuously and actively maintain contact with it.

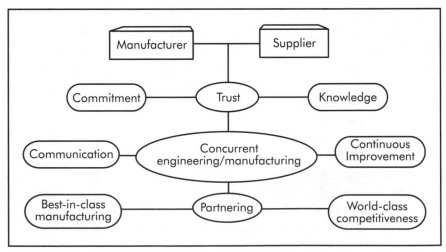

*Figure 7-2. The pyramid of business relationships.*

- The supplier has a stable management system, which is well respected by others.
- The supplier maintains high technical standards and has the capability of dealing with future technologies.
- The supplier can provide precisely those raw materials and parts required by the OEM, and meet the required quality specifications. The supplier also possesses process capabilities or has the ability to modify and enhance existing process capabilities.
- The supplier has the ability to control production volumes or to invest in equipment to meet the required production volume.
- There is no danger of the supplier breaching corporate secrets.
- The price is right and the date of delivery is met precisely. In addition, the supplier is easily accessible in terms of transportation and communication.
- The supplier is sincere in implementing contract provisions.
- The supplier is committed to cost reduction. The approach should be reasonable cost savings through process improvements or advanced technology, not through the suppliers' commitment to cut into their profit margins. Since an OEM/supplier relationship is usually for the long term, price reductions can only be realized through process-improving technology.

## Full-service Supply

An interesting offshoot of outsourcing is the so-called commodity manager or full-service supplier. The idea is to have an outside company embrace the logistics, management, and administration of in-house products needed to manufacture parts and subassemblies. The majority of the products involve tooling. Considering that the majority of manufacturing involves labor and equipment, it invites outsourcing.

The OEM wants to eliminate the management of tool acquisition, inventory control, the logistics of just-in-time (JIT) delivery, and managing the tools in the tool crib (storage, handling, setting, etc.). With this objective in mind, the OEM may plan for a short-term fix. However, this arrangement only falls far short of what is needed and might eventually lead to a dead-end relationship with the supplier.

In Figure 7-3, the left side depicts the tasks typically handled by suppliers currently in place. The right side shows the aspects that must be included in advanced OEM/supplier arrangements. The OEM needs a supplier with similar capabilities to make both parties thrive on the program. Manufacturers producing advanced, precision products would want their full-service supplier to be at the forefront of their technology and possibly have their own tool manufacturing in place. A supplier, who builds tools, certainly is more in tune with manufacturing activities and can work with its own tool suppliers sensibly to arrive at the most desirable products for its customer.

For quality- and technology-sensitive OEMs who operate in a highly competitive market, the supplier would provide extensive research and development work that would be dedicated to the needs of its OEM customer. The enabling vehicle would be concurrent engineering and, possibly, benchmarking to improve upon what is best in the industry. Performing advanced machining and manufacturing requires knowledge of adjacent fields and disciplines. For example, a cutting-tool supplier needs to be familiar with the entire envelope of the machining/manufacturing process. As such, the ability to support the production floor and offer alternative manufacturing processes should be a characteristic of the full-service supplier. The complete envelope of a full-service supplier for cutting tools should be planned. The logistics include the tasks shown in Figure 7-4.

## Knowledge, Innovation, and Technology

Manufacturing enterprises devoted to high quality, continuous improvement, and technological leadership need supply partners congruent with the same principles. It makes for a strong bond and secures continued success.

Knowledge is about learning. The process of team learning between manufacturer and supplier can lead to innovative ideas, which is necessary for the supplier to be of value to the manufacturer. The latter might need product innovation to keep up with a leadership position. The path to innovative products and processes

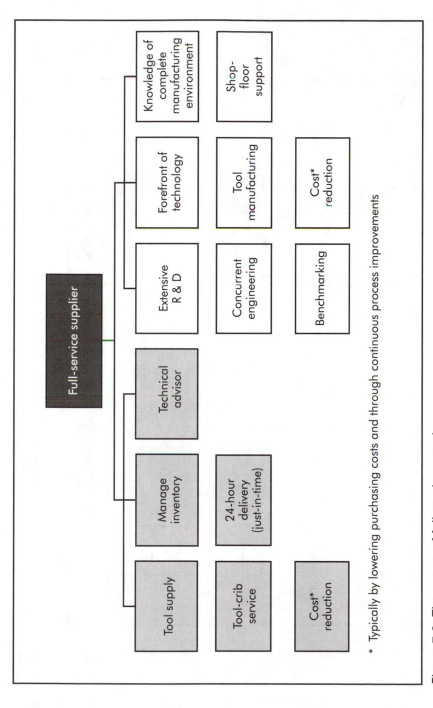

*Figure 7-3. The scope of full-service supply.*

* Typically by lowering purchasing costs and through continuous process improvements

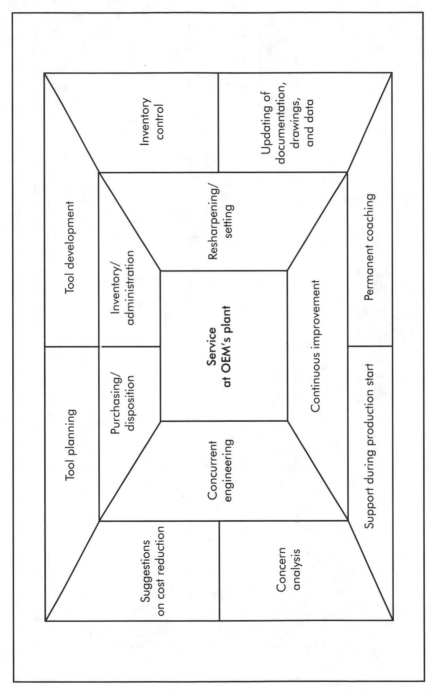

Figure 7-4. The logistics of full-service supply.

leads to continuous improvement efforts (concurrent engineering plus research and development). The supplier and manufacturer have a strong interest in this process.

Manufacturers and technology suppliers can share research and development (R&D) efforts to reduce cost and development time. This can also result in an increase of each other's R&D competence. The outcome can yield up-to-date technology way in advance of other competitors. A well-orchestrated, value-added supplier/manufacturer relationship will be the yardstick to be measured by in a market where partnerships and technology have become critical to long-term prosperity. Advanced technology, expert knowledge, customer satisfaction, concurrent engineering, innovative processes, and business excellence go full circle for the corporate strategy involving outsourcing and full-service supply, as shown in Figure 7-5.

## SPECIFYING THE BEST MACHINING PROCESSES

The knowledge of advanced technology and converting it into the most productive and economical machining process is the essence of high-performance machining. Whether the determination and layout of processes is done in-house or by a third party, it has to be done with authority and meticulously. With the advent of the full-service supplier, the increase in part complexity, and the trend to more outsourcing, expert teams must be formed to do intelligent processing.

Conventional wisdom in the automotive industry dictates that major car manufacturers should only produce between 40–45% of parts in-house and outsource the rest. A high percentage of those outsourced are technology sensitive. Persistent problems and frequent recalls often point to components made by some other company. If the problems are not assembly related, they originate in manufacturing, usually pointing to faults within machining processes. Prevention of costly repairs, rework, and recalls start with the robustness of processes. Processes have to be designed and developed with high accuracy, dependability, repeatability, and simplicity. All players and partners of manufacturing have to take ownership of how to produce the parts. The responsibility is given

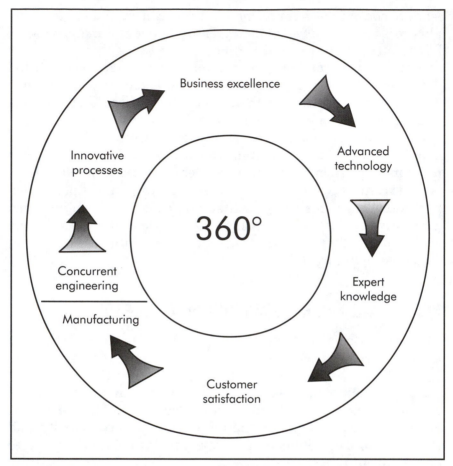

*Figure 7-5. Enablers of outsourcing and full-service supply chains.*

to the machine-tool builder, the full-service supplier of the cutting-tool manufacturer, or other independent consultants and service providers who must take ownership of the entire machining process. It must be a feasible and sensible technology-based undertaking.

Once the fundamental decision has been made as to whether to use flexible or dedicated machines, to a great extent determining the process is a matter of how, when, where, and what cutting tools to use. Therein lies the key of best processes.

The cost for tooling, as a percentage of total production, has climbed 25–30% during the past few years because of the use of flexible manufacturing. This directly influences the other 70–75% of the production cost and is why proper tool-planning layouts and specifications are so relevant. The right tooling, and not the machine, often makes the difference in precision and productivity on the production floor. The traditional approach to specifying the most-advanced machine tools, and equipping them with outdated tooling must be a phenomenon of the past. All links of the manufacturing chain have to be commensurate with one another's technology. Modern machining processes must consider and possess:

- one-pass finish-machining,
- multi-function tooling,
- no post-machining processes,
- tools made for regular production machines,
- modular design,
- standardization,
- high repeatability and accuracy,
- achievable part finishes,
- high cutting speeds and feeds,
- easy and reliable tool clamping,
- cutting material with the best price and performance,
- minimizing machining time,
- minimizing tool setting,
- defined tool management, and
- predictability of machining results.

Developing the sequential operations to finish-machine quality parts has to be done against the background of both productivity and precision. Often, special tooling has to be specified for low-batch production and new products. As the production process matures, tooling becomes more customized and fine-tuned. This results in standardization.

Today, manufacturing companies sometimes enjoy relationships with their otherwise fierce competition. This holds true for marketing, R&D, distribution, and sometimes production functions. For example, when General Motors awards a new engine program to one or two machine-tool builders, they, in turn, have a selected group or single source develop the process with them concurrently. At that point, it is crucial to specify the most-advanced tooling methods and systems, regardless of their manufactured origin.

## Making the Decision

Computer technology and software programs, the collection of past and similar machining activities, and new advancements can determine the ideal process. Regardless of brand and manufacturer chosen, the judgement is based strictly on the requirements of cost, quality, and time. High-performance machining then can indeed be expected (see Figure 7-6).

The database used to formulate the machining sequences must be updated as information on continuous improvement efforts becomes available. Manufacturing professionals must, without a doubt, stay very close to current and future technology trends and innovations. Intelligent decision-making involves determining answers to questions such as:

- What are the technological limits?
- How can we compete?
- Where do we stand in the market?
- What field would be appropriate for new technology?
- How can we increase the value added?
- What will the development cost be?
- Should we purchase an innovative idea and convert it to suitable use?
- At what point should we substitute existing technology?
- Do we want to be the low-price leader?
- Do we want to be a technology follower?
- At what point do we invest to obtain new technology?
- Can we afford to compete head-on in price or technology?

## Top Management Involvement

The next item to address is the involvement of top management since the quest for high-performance machining must trickle down the enterprise from the top. Top management business planning embraces:

- a macro overview of technological development in the industry and its adoption;
- benchmarking against the competition and adjustment of the company's capabilities in response to the results, including possible modification of R&D activities;

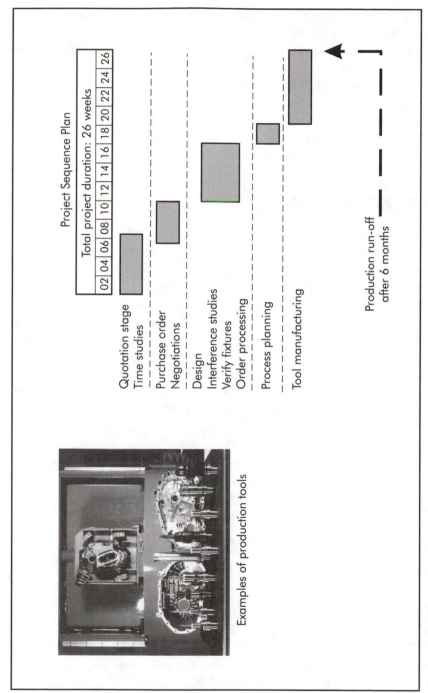

Figure 7-6. Project timeline.

- the assimilation of outside technology in response to time and cost constraints;
- connections to technology think tanks, and technology task forces in industry, academia, and government;
- closeness to customers to be up-to-date on what technology is desired;
- strong relationships with suppliers;
- technology and innovation transfer into engineering and processes, resulting in the best machining operations.

## TOOL MANAGEMENT

The instant availability of good tooling is what is traditionally expected from the tool crib. The role tool management plays in the value-adding process on the modern production floor goes beyond this expectation. As a matter of fact, it is here where the manufacturing enterprise can tip the scale for more productivity, cost savings, and continuous improvement. A thorough analysis of tool management reveals its importance for high-performance machining. Tool management is so pivotal that it must be integrated into the machining process. Tooling management comprises purchasing/disposition, inventory and control, administration, fine-tuning, continuous improvement, and inspection, as shown in Figure 7-7.

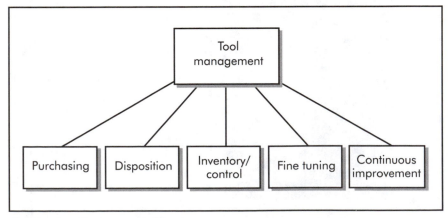

Figure 7-7. The scope of tool management.

## Purchasing and Disposition

Purchasing is concerned with the determination of order points and the supplier relationship. Disposition is concerned with distributing the right quantities to the right places at the right time. Incoming inspection should be part of the effort too, since tool-management personnel must deliver feedback on how the tools perform.

## Inventory and Control

Inventory should be lean but sufficient. This requires the compilation of all basic, pertinent information into a database. Data on use patterns, tool life, inventory levels, and costs, in addition to distinguishing characteristics for each tool should be captured. Regardless of the number of tools to manage, inventory and inventory control should be handled by an automatic inventory system, so that all activities are handled through computerized workstations (see Figure 7-8). Software packages provide for easy inventory tracking and logistics.

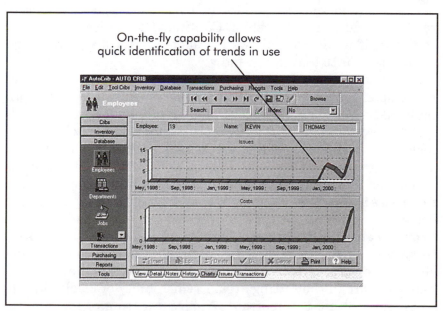

*Figure 7-8. Software depiction of tool use.* (Courtesy Auto Crib, Inc.)

## Fine-tuning

A tool is given its final touches proactively, sometimes reactively, when the information at the end of the machining process indicates that adjustments have to be made. Fine-tuning might be a final, short grind of a tool with fixed cutting edges. It could be double-checking the gage length of a tool held in shrink-fit chucks, a quick visual check for dimensions and correctness, or a functional test.

High-precision machining often means conforming to precise settings. For example, imagine a multi-function tool, performing several bore operations and facing in one pass. All step lengths and diameters have to be fine-tuned to one another. Often, the required tolerances are one step different or the tool life for the individual steps varies. This all has to be taken into account.

Another important aspect of fine-tuning is to only use the tolerances needed. Adjustable inserts can be raised or lowered from a mean diameter to stay at the upper or lower limit of the tolerance band. Lowering the insert to the low limit can add to the stability of the system. Raising it can increase tool life by allowing for more insert wear. Fine-tuning is usually done only once per system and spindle, due to the expected repeatability accuracy of the cutting-tool system. Continued proactive fine-tuning can work if production allows for adequate feedback from machine operator to tool setter and vice versa.

Accurate setting does not permit being off by even one or two micrometers. Machining, in turn, has to make use of the optimum machining data. Both depend upon proper design of the cutting-tool system (cutting geometry, system rigidity, etc.) and the machining periphery (coolant passages, chip disposal, etc.).

Manufacturing must verify that the workpiece completed at the end of the machining process meets the required geometry and finishes. Any work-in-process taken away from manufacturing for corrective action is costly. However, human error and machine defaults cannot be ruled out. Therefore, it is absolutely necessary to have two or three sets of cutting-tool systems ready for machining: one currently in use, one as backup next to the machine, and one more in the setting area. Predictability of machining results and tool life is important. Accurate setting of tooling systems with high repeatability allows for predicting tool life. Proactive fine-tuning is a necessity in high-precision manufacturing to prevent possible rework or scrap at the end of the line.

## Continuous Improvement

Continuous improvement within tool management is a necessary commitment. It calls for communication, standardization, cost savings, as well process optimization. Of these, communication and standardization deserve special attention.

### Communication

Manufacturing must be fluid, adaptive, and learn when to modify an existing process or adopt a new one. It must also learn to communicate improvements and pitfalls throughout the shop-floor organization. Many good ideas are stifled because of lack of communication. Communication can be formal or informal, depending on the subject. There should be ongoing data and idea interchange.

### Standardization

Simplicity of tool management has everything to do with standardization of tool setups and tool setting. To this end, the widespread adoption of International Organization for Standardization (ISO) quality standards has promoted uniformity in gaging equipment and processes. It has narrowed down the variability of setting and gaging equipment. This is extremely important due to the expanded supply chain, involving numerous manufacturers working on one and the same end product (a car, for example).

Only standardized, ISO-approved and maintained tool-setting and measuring equipment can identify and communicate data of meaning and consequence. Data from one source should not have to be verified by another; it should be trusted as it comes from the source. Effective tool-management systems:

- reduce overall production costs;
- increase supplier involvement and responsibility;
- attain quality goals;
- facilitate manufacturing;

- manage technology changes;
- enhance the production floor's team structure;
- improve coordination between engineering and manufacturing;
- pursue doing it right the first time;
- provide a valuable database; and
- minimize process complexity (Hronec 1993).

## REFERENCES

Chopra, Sanil and Peter Meindl. 2000. *Supply Chain Management: Strategy, Planning, and Operations*. Englewood Cliffs, NJ: Prentice Hall.

Hronec, Steven M. 1993. *Vital Signs*. New York: Arthur Anderson & Co.

# Dry and Near-dry Machining

8

This chapter will discuss dry- and near-dry machining, its environmental impact, and principles for application. It will also discuss how to reduce coolant use, by using biodegradable and recyclable coolants.

## ENVIRONMENTAL IMPACT

Environmental impact was not an issue in the industrial revolution or even during the post-war era. However, it is slowly being recognized as an important factor for designing an efficient operation. The process of making products has to be done with the least possible harm to the environment. Reduction of energy usage, substitutions for traditional material, different designs, applying new advanced machining technology, and promoting the use of recyclable and biodegradable materials can have enormous impacts on securing a protected, healthy environment. Some governmental laws and restrictions are already in place. Behind the challenges ahead lie untapped opportunities for new technology, new product lines, and new markets. Environmental considerations must become part of the enterprise.

### Coolant Use

Every year, manufacturing consumes a huge amount of coolants. While they aid in the outcome of the machining process, they are costly and pose a threat to health and the environment. Health experts and environmental groups point out that coolants contaminate our soil (leakage), water (disposal), and air (airborne particles), and could cause skin and lung diseases. Recent studies found that the cost for coolants is 7–16% of the total production cost. This includes the cost for their acquisition, maintenance, and disposal. These issues have compelled manufacturing companies to drastically reduce coolant consumption and, if possible, eliminate it altogether.

Just turning off the flow of coolant does not eliminate its effects (see Figure 8-1). In fact, the use of wet machining may never be abolished. Non-geometrically defined machining processes are very coolant sensitive. In this case, the plan should be to exercise coolant restraint, optimize consumption, and more importantly, use substitute processes. Machining without coolants is promising for geometrically defined machining processes, which include the majority of operations (see Figure 8-2). However, it is essential that every machining parameter be considered.

## DRY MACHINING

Despite the drawbacks of using coolants, they have three vital functions in the machining process:

- to cool;
- to lubricate; and
- to discharge the cutting chips.

Eliminating coolants and, with them, their primary functions of cooling, lubricating, and chip flushing, has consequences for the workpiece, the cutting-tool material, and the machine. Not only that, the process takes place at much higher friction and with more adhesion between the part and the tool. The heat generated and transferred to the workpiece, cutting tool, and chip is no longer absorbed by the coolant; it is taken out of the immediate cutting area. Heat buildup affects the cutting edge of the tool. Lack of lubrication affects the surface finish. And, improper discharge of cutting chips detrimentally affects both the workpiece and the tool.

The most challenging machining/material combinations are shown in Table 8-1. With the exception of machining superalloys, turning and milling can be done without coolant assistance.

Removing coolants from the machining process requires special design features for the machine tools. The two areas affecting

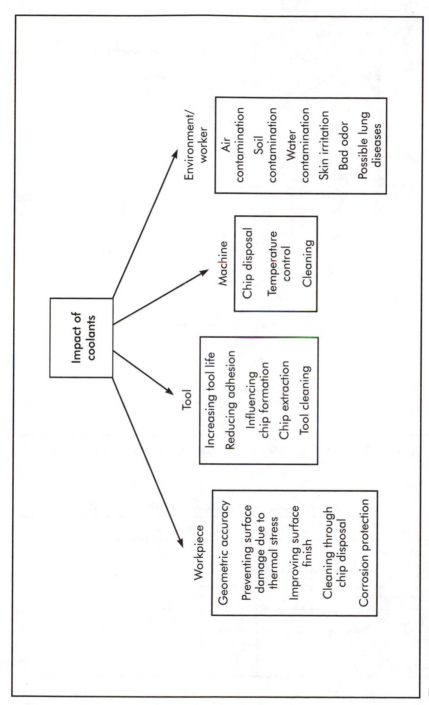

Figure 8-1. Impact of coolants.

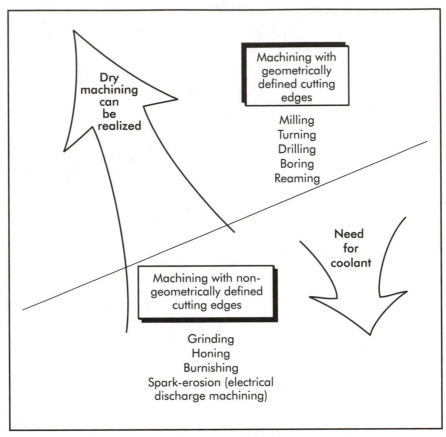

*Figure 8-2. Machining operations and their need for coolant.*

Table 8-1. Most challenging machining/material combinations

| Workpiece Material | Turning | Milling | Reaming | Threading | Drilling |
|---|---|---|---|---|---|
| Cast iron | | | | | |
| Steel alloys | | | * | * | * |
| Hardened steel | | | * | * | * |
| Aluminum | | | * | | * |
| Superalloys | * | * | * | * | * |
| Composites | | | | | |

the machine are thermal stress and chip control. Existing machines have to be modified and new ones designed accordingly right from the start. Special design features for machines are few and the conversion from wet to dry machining requires only a few modifications. Once they are part of the machine's design, dry or near-dry machining or conventional machining, can be done. Figure 8-3 shows the few necessary special design characteristics for the machine.

## Workpiece

The characteristics of the part to be machined determine whether or not it can be manufactured in a wet or dry process. The quality of the finished part is one of the key outcomes of the machining process.

The objective is to transfer as little heat as possible to the part. It has been proven that a great part mass does not warm up easily. Therefore, parts of greater mass are more suited for dry machining. Near-net-shaped parts can be produced in a single pass.

From a workpiece standpoint, the following should be considered when machining dry:

- increase metal-removal rate per time unit;
- use of cutting force reducing tool geometry;
- workpiece density and heat conductivity;
- the most suitable machining operation; and
- tool design and machining data.

## Cutting-tool Design

The focus for optimizing tool design is on cutting geometry and chip removal. A reduction of the cutting force through a wider clearance angle yields less tool stress and lowers the cutting temperature. Defined rake angles aid in breaking up the chips. Wide chip galleys and special flute design (drilling and threading) assure fast and easy exit of the cutting chips.

### Material

Dry machining requires cutting-tool materials with extreme hardness as well as abrasion, thermal shock, and adhesion resistance.

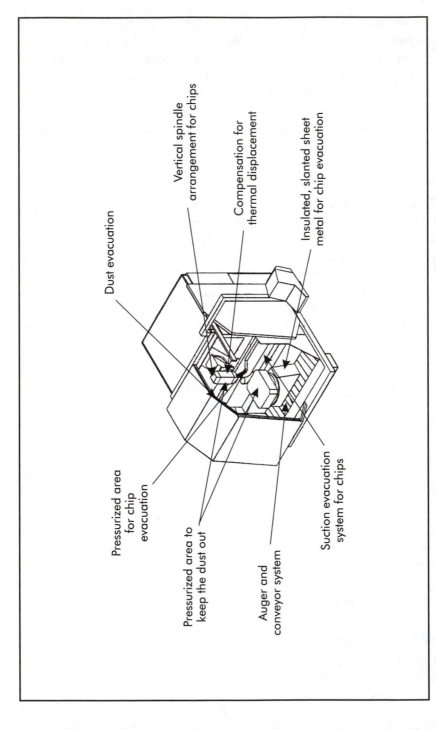

*Figure 8-3. Special design features for dry machining.* (Courtesy Fraunhofer Institute for Production Technology)

The most suitable materials are cubic-boron nitride, polycrystalline diamond (PCD), ceramics, cermets, and some diamond-like coatings.

Tightly toleranced bores are often machined with tools featuring guide pads. PCD guide pads prevent chip build-up because of their non-affinity to most workpiece materials. Their hardness lowers friction and thus heat buildup. PCD guide pads are highly suited for dry machining.

The National Center for Manufacturing Sciences (NCMS) extensively tested dry-hole drilling of aluminum and found physical vapor deposited, diamond-like coating best when compared to any other coating in terms of tool life, as shown in Figure 8-4. This same study, sponsored by the Big-Three car manufacturers, also found that the dry machining particle count and size, which are similar to those in wet machining, fell well below the partial-exposure limit, as shown in Figure 8-5. Therefore, no health and safety hazards are posed.

## Machining Data

High cutting speeds are generally desirable for dry machining because the cutting tool and workpiece have less contact during cutting. The result is lower tangential cutting force and workpiece temperature. The lower the workpiece temperature, the less surface distortion and the more likely that parts are finished to blueprint specification.

Even high feed rates are desirable, for the cutting heat is not absorbed by the workpiece, but rather by the cutting chips. The only problem is that higher feed rates call for higher cutting forces, but that can be dealt with through proper tool design, for example, by using tangential inserts.

## NEAR-DRY MACHINING

When dry machining is not feasible technologically, particularly for drilling, reaming, and fine boring, near-dry machining is an alternative. The goal is to induce a metered, minimum amount of lubricant onto the cutting edge. This principle of minimum-volume lubrication replaces conventional flooding during machining; it is, in fact, a near-dry process.

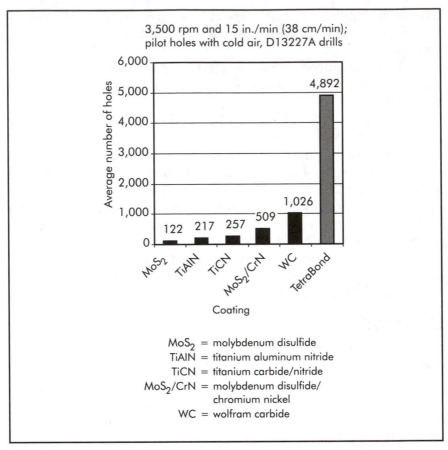

*Figure 8-4. Comparison of drill-point coatings.*

Table 8-2 provides examples of part materials processed by precision fine boring with minimum-volume lubrication.

## Principles for Application

Typically, machining centers handle coolant volumes of 5–53 gal/min (20–200 L/min). With minimum-volume lubrication, only 4 oz/hr (0.13 L/hr) to a maximum of 7 oz/hr (0.2 L/hr) of fluid is used. The vegetable-oil based lubricants are nontoxic and biodegradable. Their application is safe, controllable, precise, and can be monitored with devices that automatically check temperature and dimensional part deviation from the blueprint.

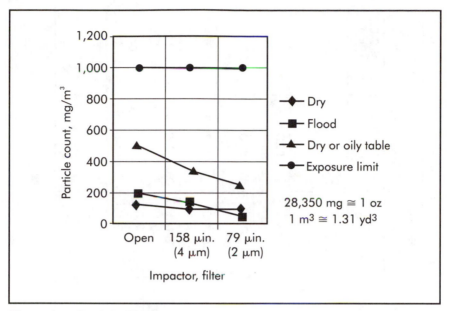

*Figure 8-5. Particle filter.*

Small lubrication particles are mixed with air and directed to the cutting tool either from outside the machine or through the machine spindle. Mixing with air requires no major modifications of the machine. Its disadvantage is the length-to-diameter ratio of about 2–2.5, since no lubrication can be brought to the cutting edge when machining bores. For the through-to-spindle lubrication supply, it is necessary to design the spindle for dry rotation. This method offers a controlled and versatile process. The oil/air mix can be induced into the tool on the cutting edge primarily through the four different ways shown in Figure 8-6. These methods ultimately depend on machine layout, part design, and workpiece material.

## Cooling and Lubricating

Of the two functions of cooling and lubricating, cooling plays the minor role in minimum-volume lubrication. This is because of the hot hardness of advanced cutting materials. Also, temperatures transferred to the workpieces can be kept to a minimum with the right cutting parameters. Minimum-volume lubrication

**Table 8-2. Examples of materials processed by precision fine boring with minimum-volume lubrication**

| | Grey Cast Iron ASTM/A48, CL 40 | Pure Iron | Pure Aluminum AISI/422 | Stainless Steel | Steel ASTM/1060 | Grey Cast Iron |
|---|---|---|---|---|---|---|
| Bore diameter, in. (mm) | 5.1 × 5.1 (130 × 130) | 1.4 × 2.4 (35 × 60) | 0.8 × 2.4 (20 × 60) | 0.8 × 2.0 (20 × 50) | 0.8 × 2.0 (20 × 50) | 0.8 × 2.0 (20 × 50) |
| Tolerance | K6 | P7 | H6 | — | — | — |
| Target, in. (mm) | Mean | — | 0.7876–0.7878 (20.006–20.009) | | | |
| Guide-pad material | Polycrystalline diamond (PCD) | Cermet | PCD | PCD-cermet | PCD-cermet | PCD-cermet |
| Insert material | Cermet | Cermet | PCD | PCD-cermet | PCD-cermet | PCD-cermet |
| Cutting speed, ft/min (m/min) | 1,312 (400) | 820 (250) | 2,625 (800) | 656–722 (200–220) | 656 (200) | 410–574 (125–175) |
| Feed/rev., in. (mm) | 0.003–0.004 (0.08–0.10) | 0.004 (0.10) | 0.004 (0.10) | 0.002–0.005 (0.050–0.125) | 0.0010–0.0040 (0.025–0.100) | 0.002–0.004 (0.05–0.10) |
| Depth of cut, in. (mm) | 0.01–0.02 (0.3–0.5) | 0.006 (0.15) | 0.004 (0.10) | 0.004–0.006 (0.10–0.15) | 0.004–0.006 (0.10–0.15) | 0.004 (0.10) |
| Insert geometry Radial rake angle Lead angle(s) | Hexagonal 12° 3°/30° | 12° 3°/30° | 6° 75° | 8° 3°/30° | 12° 3°/30° | 6° 3°/30° |

## Table 8-2. (continued)

| | Grey Cast Iron ASTM/A48, CL 40 | Pure Iron | Pure Aluminum AISI/422 | Stainless Steel | Steel ASTM/1060 | Grey Cast Iron |
|---|---|---|---|---|---|---|
| Surface finish, Rz, mil (μm) | 0.08–0.2 (2–5) | <0.2 (<5) | <0.08 (<2) | <0.04 (<1) | <0.06 (<1.5) | <0.06 (<1.5) |
| Bore geometry, circularity, cylindricity, mil (μm) | 0.16–0.28/ <0.16/<0.24/ (4–7/<4/<6) | <0.16/<0.12/ <0.08 (<4/<3/<2) | <0.06/<0.08/ <0.08 (<1.5/<2/<2) | <0.08/—/ <0.12 (<2/—/<3) | <0.08/—/ <0.12 (<2/—/<3) | <0.08/—/ <0.12 (<2/—/<3) |
| Coolant usage | <0.034 oz/hr (<1 ml/hour) | — | <0.034 oz/hr (1 ml/hour) | <0.01 oz/ bore (<0.3 ml/ bore) | <0.01 oz/ bore (<0.3 ml/ bore) | <0.01 oz/ bore (<0.3 ml/ bore) |
| Coolant introduced | External | Central | Central | Central | Central | Central |
| Air pressure, bar | 6 | 6 | 6 | 6 | 6 | 6 |

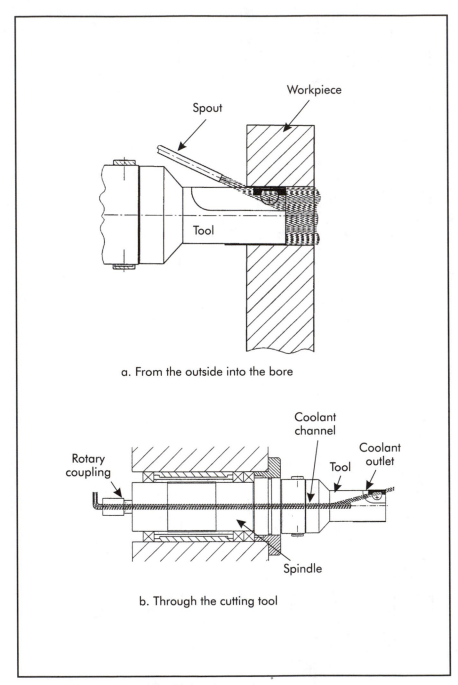

a. From the outside into the bore

b. Through the cutting tool

*Figure 8-6. Inducing the oil/air mix.*

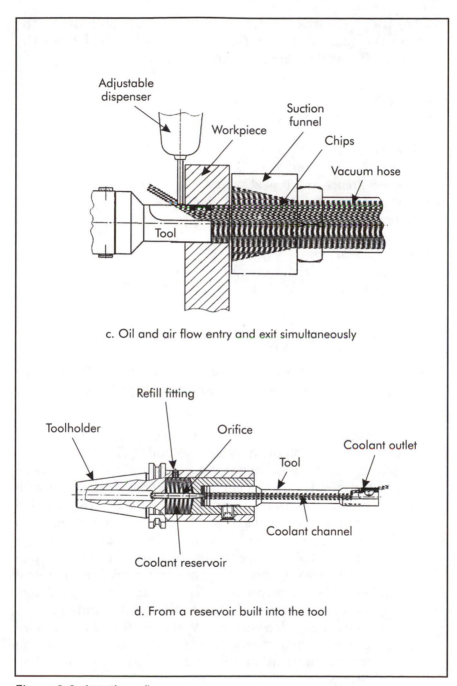

c. Oil and air flow entry and exit simultaneously

d. From a reservoir built into the tool

*Figure 8-6. (continued).*

reduces friction, curbs the cutting temperature, and lowers the wear on the tool. The lube film separates the tool from the workpiece effectively, which prevents adhesion.

## Advantages

Minimum-volume lubrication offers many advantages:

- low cost,
- no coolant maintenance,
- the workpiece and tool stay dry,
- can be installed into existing machines,
- minimal energy consumption,
- vegetable oil is biodegradable,
- no toxic additives,
- no known skin irritation, and
- produces dry chips to sell.

## REDUCING COOLANT USE

There are some options to substitute for cutting fluids. Surprisingly, they are not well known and are, therefore, not taken advantage of:

- Use cutting-tool geometry that reduces cutting forces and edge geometry with primary and secondary lead angles.
- Specify a workpiece of high-density material to better absorb the heat generated during cutting. Or, choose a material with high-temperature conductivity, so the heat is easily carried away from the machined surface. Heat should not be allowed to detrimentally affect the workpiece (for example, warping).
- Design cutting tools with cutting edges of extreme hardness and adhesion resistance, particularly polycrystalline diamond (PCD) and cubic-boron nitride (CBN). Ceramics are good, too, except they are vulnerable when making interrupted cuts.
- Use physical vapor deposition (PVD) titanium aluminum nitride (TiAlN) or molybdenum disulfide ($MoS_2$) coatings on tools for their lubrication effects and, if possible, multi-layered coatings.

## Stock Removal

The strive for more near-net-shaped material works well for near-dry machining in that the depth of cut can be kept to a minimum. Lighter cuts cause less heat and are, therefore, preferred. However, it is better to have chips than none at all, especially when machining at higher feeds and speeds. The heat escapes into the chips rather than staying at the immediate point of cutting.

## Speeds and Feeds

High cutting speeds generally are desirable for dry machining. They lower the tangential cutting force and reduce workpiece temperature, because the cutting tool and workpiece are less in contact during cutting. The lower the workpiece temperature, the less surface distortion, and the more likely parts are finished to blueprint specifications. High feed rates shorten the cutting length to minimize the cutting heat absorbed by the workpiece. The only problem is that higher feed rates cause higher cutting forces. This might cause more stress to the tool and machine spindle.

## Automotive Industry Takes the Lead

A study conducted by Ford Motor Company in 1997 showed that four of their main U.S. transmission plants produced roughly three million automatic transmissions. For this production, they used 5.6 million gallons (21,198,307 L) of coolant. In 2000, Mazda reported that 30% of the energy consumed within their machining lines was related to the central coolant system. They further stated that the cost for coolant disposal was almost half of their entire day-to-day production costs. In addition, about 50% of the original coolant volume was lost through splash and evaporation during machining.

General Motors, Ford Motor Company, DaimlerChrysler, and many of their key suppliers have launched extensive programs toward eliminating, or at least curbing, the consumption of coolants. First attempts toward dry or near-dry machining for higher-volume production are promising.

Generally, coolants are still an important part of most chip-making processes. Unquestionably, its cooling and lubrication inhibitors positively affect the quality of workpieces, prolong tool life, and assist in chip discharge. Today, it is estimated that the Big Three automakers machine 85–90% of their workpieces using coolant.

Occupational Safety and Health Administration and Environmental Protection Agency requirements, as well as cost-cutting efforts are the right incentives to pursue technology for coolant reduction. The path to more controlled coolant usage in automotive manufacturing will lead to curbing consumption. In the case of stand-alone machines, coolant reduction is easier to realize than with transfer lines, where sequential machining operations are done and all coolant is supplied from a central coolant reservoir. A good start is to direct coolant only to the stations where it is absolutely needed, monitor its flow through reduction valves, and eliminate coolant when the line is down. The overwhelming trend toward flexible machining centers will help reduce coolant use.

Considering the technical expertise and technological advancements available, it is prudent to explore and pursue other avenues. Wet machining may not be entirely eliminated because some machining operations will always need some form of traditional cutting fluids, at least in the foreseeable future. However, first attempts at dry or near-dry machining have proven its feasibility at low cost and a minimum of technical effort. Technological advancements in tooling, the use of substitute machining processes, as well as the development of substitute coolant media make it possible.

Reductions in metalworking fluid use can contribute to more efficiency and productivity on the production floor. Advancements in machining technology have opened up new avenues. Retrofitting machines for dry or near-dry machining is relatively inexpensive. New machines should incorporate the design features needed for it. Fine-tuned cutting-tool technology will guarantee its success.

## Biodegradability and Recyclability

Tailor-made coolant types and grades for different machining operations add complexity to the shop floor, which make maintenance and disposal not only more expensive, but can increase contamination risks to the air, soil, and water. For example, the use of different fluids for grinding, honing, milling, and drilling make the processes more complicated. The first step should be to specify the same cutting fluid for all operations, specifically a type with low emission grades and biodegradability.

Biodegradable oils can contribute to cost savings in two areas: recyclability of up to approximately 25% and increased overall tool life. Thorough drying of workpiece chips enables reuse for casting new workpieces. To economically recycle products at the end of their life cycle with the least amount of effort, the principle of design for disassembly and recyclability must be applied during the design stage. In fact, all objectives for a better environment should include recyclability, which includes weight and dimensional reductions, volume curbing, higher efficiencies, less emissions, etc. For example, when the steel industry realized how important weight reduction was for the automotive industry, they developed thinner steel sheets that still maintained their original physical characteristics. This enabled them to compete with suppliers of plastic materials. Because steel had better recycling capabilities, auto manufacturers were then inclined to continue steel usage.

## CONCLUSION

Manufacturing enterprises have long seen environmental issues as just another burden, which is largely due to the legislative measures taken to mandate producing products with less pollution and with more regard for health. Today, corporations realize that while reducing waste and lowering emissions is a challenge, the process can also open doors to new opportunities and substantial benefits.

Ultimately, if environmental considerations are to become an integral part of the entire manufacturing enterprise and not just isolated functions of after-thoughts, revisions, and temporary fixes, the following items need to be implemented:

- Develop a positive and active approach to health and pollution issues throughout the organization.
- Design accommodations for the disposal of manufacturing process waste.

- Establish environmental standards.
- Management must realize that employees' health is human capital.
- Create environmentally friendly processes and products that secure future growth.

In practice, many manufacturing companies still take corrective environmental measures only when mandated. However, as integration becomes part of corporate culture, companies will embark on environmentally oriented manufacturing processes and, ultimately, move on to the end products. The need to improve processes and products can lead to higher degrees of innovation and leapfrogging technology. This can equally open up or even create new markets and, as a result, yield high return on investment and secure corporate success.

## REFERENCE

Erdel, Dr. Bert P. 1999. *Environmental Issues in Machining*. Technical paper. Dearborn, MI: Society of Manufacturing Engineers.

# The Transition to High-performance Machining

# 9

Sustained high-performance machining can only be secured by a company that embraces it throughout its organization, making it part of all processes, principles, and systems. This goes beyond equipment, machinery, and tooling. The goal is to become a winning company through the pursuit of innovative technology and processes. Past success should not be the limit of the future.

## TECHNOLOGY FORECASTING

Technological progress in just the past five years has been dramatic. The Internet has exploded around the globe. Gene splicing has set the biotechnology industry in accelerated motion, genetic engineering is transforming medicine and agriculture, and faster computers have spurred manufacturing's productivity. Information technology is transforming all kinds of industries such as finance, business services, entertainment, and communication. Developments in micro-electro-mechanical systems (MEMS) will make the incorporation of tiny sensors, motors, and pumps into microprocessors possible, and revolutionize designs in automotive, aircraft, and appliance industries. Scientists are learning how to create materials atom-by-atom, potentially transforming the entire world of manufacturing.

Of course, it is difficult to predict which innovation will succeed, what technology to bring to market, and what customers will accept. For any technology-based enterprise, it is important to be aware of potential innovative developments and opportunities. Forecasting advances in technology and their impact on the market must be part of a company's strategic planning.

Technological leaders maintain their leadership role by scanning their own and other markets for innovation that will steer research and development (R&D), pricing, distribution, procurement, and possible mergers and acquisitions in the right direction. Knowledge of industry-wide developments helps in making decisions to fend off competition or adjust to potential threats. Technological changes can redirect entire industries, threatening any company that is wrongly positioned.

*Technological forecasting* is defined as "the description of prediction of a foreseeable technological innovation, specific scientific refinement, or likely scientific discovery, which promises to serve some useful function, with some indication of the most probable time of occurrence."

There are defined methods that can be applied empirically and mathematically for technology forecasting. The major ones are:

- trend interpolation—using trends to project into the future;
- delphi—committee meetings of experts;
- scenarios—think tanks collecting data supported by extensive research work;
- relevance trees—evaluating alternative paths;
- cross-impact analysis—recognizing interrelationships with other areas; and
- technological substitution—describing the growth pattern of S-shaped curves.

The techniques used are a function of the industry, market type, and the size and financial strength of the enterprise. None of them can substitute for having a pulse on the market and being aware of new technological developments and their impact on present business. Top management needs to have reliable sources of information and the process of gathering data has to be organized in a systematic fashion. Information is gathered by monitoring innovative efforts in industry, adjacent sciences, and tangential technologies, including any new developments that could trigger new market demand and adjustments.

Management can assume a proactive posture by recording the information, including cost, time, performance, degree of impact, and relating it to current plans and capabilities. Forecasting technology innovation is always market oriented. Technological progress often responds to a market need. However, technology can also create markets. If there is an urgent need for new technology and a healthy return on investment for the technological innovator, then the pressure is high for financial support of the development program. The result is usually a new process or a new product.

Systematic forecasting and business planning are fundamental to making intelligent decisions. Decision-makers must:

- have a macro overview of developments in industry and monitor the pulse of the market;
- compare and adjust status and abilities in relation to competitors, and modify research and development (R&D) activities and corporate strategy;
- redirect the company's strategy to avoid undesired confrontation with competitors and potential threats to the enterprise;
- be aware of forthcoming technology and innovation and keep track of its progress to make a timely move; and
- be close to the customer and market (Dundon 2002).

## Vision Into the Near Future

As previously stressed, the most important aspect in forecasting is to see the big picture, draw sensible conclusions from it, seek realistic opportunities, and get started on the activities needed in a timely manner. A look at the automotive industry illustrates the importance of being informed, involved, and observant. See Figure 9-1.

To find opportunities outside a company's core activities, explore information on R&D underway in a particular field, or connect research with others, the National Science Foundation (NSF) is an excellent resource. So are the National Center for Manufacturing Sciences (NCMS) and the National Advanced Manufacturing Test Board (NAMTB). All three connect government projects with industry and academia.

Some interesting activities underway are:

- the inter-repeatability of manufacturing systems through a standardized infrastructure;
- information-based manufacturing;

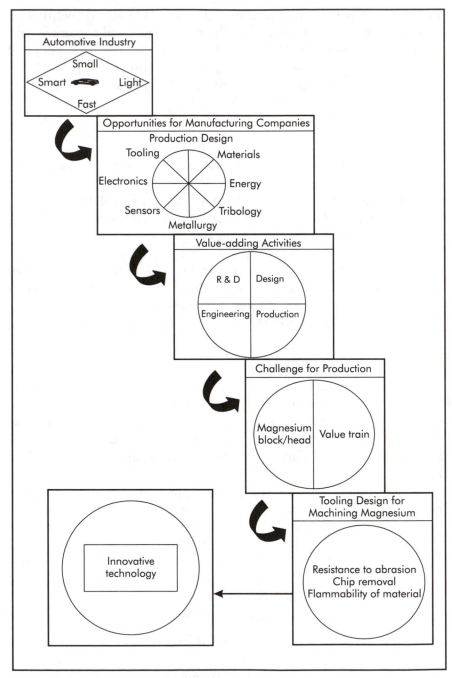

*Figure 9-1. Innovative technologies.*

- atom-based dimensional standards;
- fast modeling and simulation systems;
- full-scale rapid prototyping;
- extending the limits and performance of Stewart-platform-based machine tools (hexapods);
- lightweight artificial materials resistant to high abrasion and high temperatures;
- elimination of machine-tool errors;
- submicron machining of ferrous and nonferrous materials;
- economical, productive dry machining;
- ultrasonic, cryogenic, and laser machining;
- diamond coatings for regular production machining;
- next-generation intelligent manufacturing systems; and
- reducing time to market.

A manufacturing enterprise will have to be innovative in all facets of its business. It will need advanced technology and to effectively groom the knowledge of its staff. Pursuing knowledge, innovation, and advanced technology is the trilogy for continued future success (see Figure 9-2).

## PROCESS-ORIENTED MANUFACTURING

Typically, a given machining process follows the hyperbolic curve of interdependence between productivity and precision: an increase

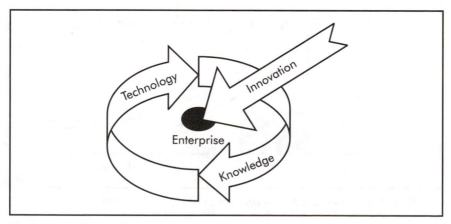

*Figure 9-2. Guarantors of success.*

in productivity occurs at the expense of precision and vice versa. In other words, they are dependent on each other within their respective parameters. By accepting a certain range of productivity and precision, manufacturing establishes comfort zones to produce parts. For example, machining performance can be measured in feed rates or stock removal. In another example, the variable for precision could be bore roundness or surface finish (texture). Changing the variable of one, changes the other. For example, when a reamer has to finish a bore with finer finish (precision), the respective feed rate (performance) has to be changed, as shown in Figure 9-3.

To improve both performance and precision without sacrificing one for the other, the process has to be improved. This might entail phasing in new manufacturing processes. Unless faced with completely outdated machines, a changeover to advanced cutting-

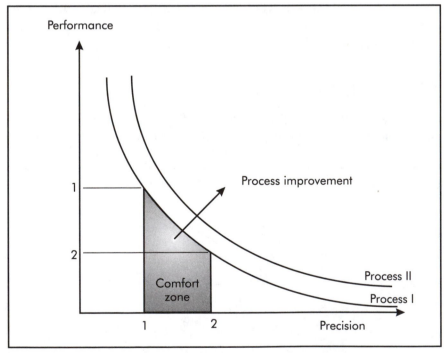

*Figure 9-3. The relationship of precision and performance.*

tool systems is prudent and sufficient. It is important that manufacturing think "process." That is, specify and monitor all technical aspects at every machining station.

Statistical process control (SPC) has helped raise the awareness of the manufacturing process. It detects variations before out-of-control conditions occur. And, process capability ($C_{pk}$) identifies how good a process is, based on the manufacturing floor's machine capabilities. Emphasizing process rather than product, these tools help determine the causes of nonconformance and lead the way toward quality manufacturing: less waste and rework and low-cost machining.

## CONCURRENT ENGINEERING

In its basic, traditional form, the organizational structure of a company can be described as isolated departments with self-contained functions. This dictates a rigid, sequential process from a product's design to its manufactured completion. Isolated and sequential work hampers the timely, interactive flow of necessary input and feedback throughout the organization. Departmental isolation and sequential, self-contained work needs to be converted into departmental integration and overlapping work modes. Concurrent engineering efforts make this possible by bringing all entities affected by product and process together at the early design stage (see Figure 9-4). The objectives of concurrent engineering are shown in Figure 9-5.

First, there must be a cohesive understanding of the technology needed to produce the product. Information sharing is accomplished through cross-functional teams from all areas of the manufacturing company, including design, production, processing, purchasing, and marketing.

Secondly, the teams need to include the customer from the very beginning. Customer feedback should be sought intermittently throughout the process.

Thirdly, concurrent engineering necessitates the involvement of relevant suppliers. The key is to bring overlapping teams together to optimize product and process.

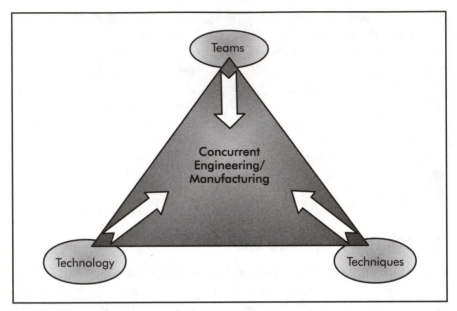

*Figure 9-4. Concurrent engineering trilogy.*

Concurrent engineering includes several techniques. They are:

- design for manufacturability (process capabilities);
- design for reliability (maintenance and life cycle);
- design to cost (product cost and marketing price);
- design to application (consumer product/production product); and
- design to quality (total quality management [TQM], statistical process control [SPC], etc.).

In manufacturing, product and process go together. The process is determined by the methods applied. A machining system is primarily composed of machine-tool systems and cutting-tool systems. All too often, concurrent engineering takes place without consulting and involving the cutting-tool system supplier. In other cases, the tooling-system manufacturer becomes involved after the final design stage, often halfway through completion of building the machine tool.

The neglect is particularly obvious when customers order turn-key systems. The machine-tool builders often leave it up to them-

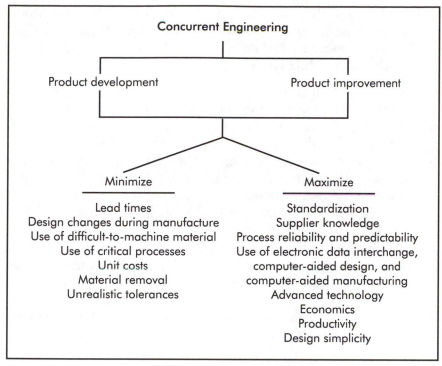

*Figure 9-5. Objectives of concurrent engineering.*

selves to source the appropriate tooling. This defeats the purpose of concurrent engineering because the cutting-tool systems constitute one of the most pivotal elements within a manufacturing process. The impact that cutting tool systems have on the outcome of the manufactured product demands their consideration right at the outset. The tooling-system supplier must be included on the team. Neglecting this is inexcusable and detrimental. As part of the team, the tooling supplier can contribute their knowledge and expertise to help with product selection.

In particular, a cutting-tool system supplier should:

- have an in-depth knowledge of the interaction of machine tool, cutting tool, and workpiece;
- be open to discuss alternate machining processes;
- offer a loop of single-source responsibility (design, manufacturing, marketing, application, service support, and training);

- understand the machining periphery, such as tool and work-piece clamping, coolant, and tool management; and
- conduct research and development to be able to offer state-of-the-art products at any time.

About 75% of all errors are made in product development, design, and planning. And, they usually do not show up until at or near the end of the manufacturing processes. A prerequisite of avoiding such pitfalls is the involvement of all relevant contributors in concurrent engineering. Otherwise, opportunities are missed and an essential tool, which is employed to improve process and product value, is lost.

When the world economy entered a new growth phase, American manufacturers suddenly were faced with a slew of new competitors. Subsequent restructuring of entire industries has evolved around several factors. One is to develop, produce, and market products of high quality more rapidly. Another is to be able to respond to erratic market changes with speed and accuracy and be closely positioned to the customer. A third is to realize that there is more complexity in terms of products and the business environment.

Manufacturing must be innovative, make technology and knowledge available, and unite product and process throughout the manufacturing enterprise. The impact on restructuring efforts in engineering and manufacturing are dramatic. Many corporations failed to adjust. They either were swallowed up by competitors or went out of business. Others tried to re-engineer or reinvent themselves without much success and had to settle for second- or third-string production, hoping to survive with their products in aftermarkets.

More complexity can only be mastered through new and advanced structures, techniques, and strategies. There is no better way to embrace product and process complexity than with the principles of concurrent engineering and concurrent manufacturing. The former has been around and applied throughout every branch of manufacturing, the latter has yet to be instilled upon corporate production executives and managers.

The principles of concurrent manufacturing are simple and direct. However, they have to be applied with discipline in a well-organized, highly open, communicative way, and in an expert team setting. To fully understand applied concurrent manufacturing is to review, comprehend, and build upon the fundamental concurrent engineering principles.

## New Manufacturing Paradigms

Product design and development has changed manufacturing forever. For example, manufacturing is now done in a smaller world market that is more interdependent, more quality minded, more cost conscientious, and more complex.

A couple of decades ago, Japanese manufacturers showed everyone how to define absolute and sustained product quality. Quality-mindedness after that was simply a matter of staying competitive with imported products. With it came unrelenting requests from customers, especially those of the major automotive manufacturers, to offer optional equipment and make cars available as specified for immediate purchase and delivery.

As the world was becoming one global marketplace, more competitors began to vie for new customers, accelerating the speed of design, development, and manufacture of new products. This, combined with growing parts and products complexity spurred by environmental, energy, and health considerations, has increased the complexity of an already multi-faceted manufacturing forum. Many companies have considered re-engineering and reinventing themselves. The goal has been to reorganize from the bottom up and redefine the purpose and vision of the corporation. Corporate culture must embrace knowledge, technology, and innovations. Agility, benchmarking, and continuous improvement are also important. The defined production goals must be optimized productivity, efficiency, and quality.

It became clear that the corporate structure, built rigidly in a vertical mode with more isolated functions and departmental self-autonomy, had to change. The barriers to better information flow and departmental overlapping had to be removed. Above all, corporations with more advanced hierarchical organizational structures began to realize the real value of teamwork. Disseminating relevant information and discussing problems increased knowledge dramatically. As pressure mounted to deliver more in less time and to combat shrinking product life cycles, manufacturers extended their meetings to team up with their customers. The intention was to collaborate early at the product design and development stages to satisfy customer needs before product manufacturing commenced. This concurrency has created new ways to manage the demands on product design, development, and manufacturing.

## CONCURRENT MANUFACTURING

The dictionary defines *concurrent* as "occurring, arising, or operating at the same time, often in relationship, conjunction, association, or cooperation." Concurrent manufacturing is as multi-faceted as this definition because it can be applied in close relationships, in tight associations, and within smart corporations. In fact, concurrent manufacturing strategies such as outsourcing, full-service supply, offshore manufacturing, joint ventures, and globalization can be embraced in different settings (see Figure 9-6).

Original equipment manufacturers (OEMs) have started to narrow down their scope of suppliers and will continue to do so to facilitate logistics, cost cutting, and better communication.

Purchasing certain families of products through a full-service supplier makes perfect sense as part of a leaner enterprise. The supplier, in turn, gets sustained, captive business, and growth. From a technological point of view, however, the mutual dependence harbors potential risks if:

- there is an imbalance of power;
- there is not enough interaction;
- not all resources are committed; and/or
- each other's quality standards are not commensurate.

Concurrent manufacturing gives the OEM and the full-service supplier the means to forego a dangerous liaison. In a team-structured setting, involving multi-functional personnel, the OEM must know and assure that its procurement represents state-of-the-art in quality and price. OEMs also must have a built-in yardstick and seek verification that they are not shut out from technological ad-

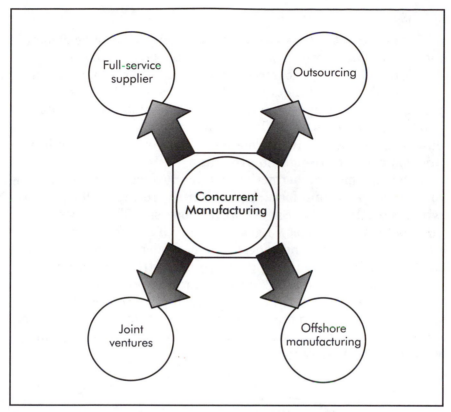

*Figure 9-6. New corporate strategies.*

vancements by their current supplier. The full-service supplier needs to know what is out there. The goal is to provide the OEM with the best available product and service, regardless whether or not it is part of their own lineup. If the full-service supplier depends on their supply from other suppliers, they also have to be involved with that supply chain. They also have to stay abreast on new developments, expectations, adjustments, etc., which are the result of the concurrent manufacturing methodology.

The principles of concurrent manufacturing are derived from concurrent engineering. The latter was first; the former will very much determine the competitiveness of any manufacturing entity. Both embrace the triangle of teams, techniques, and technology as was shown in Figure 9-4.

## Teams

Teams are at the core of concurrent engineering and concurrent manufacturing efforts. The power is in the pooling of professional backgrounds, skills, expertise, knowledge, and the application of problem-solving techniques. The team is committed to a common purpose, guided by performance goals, and made accountable for progress. Members share leadership roles, encourage open discussions, and measure performance collectively. Team commitment and competence strengthen over time.

The fundamental distinction between concurrent engineering and concurrent manufacturing is that the former is for products while the latter is concerned with processes. That is why team composition varies (see Figures 9-7 and 9-8). However, it would be an advantage for the concurrent engineering team to include manufacturing personnel and vice versa for the concurrent manufacturing team. Essentially, either team should include representatives of the relevant departments including, but not limited to, design, engineering, manufacturing, product support, purchasing, quality, marketing, production processing, and planning, and—very important—key suppliers.

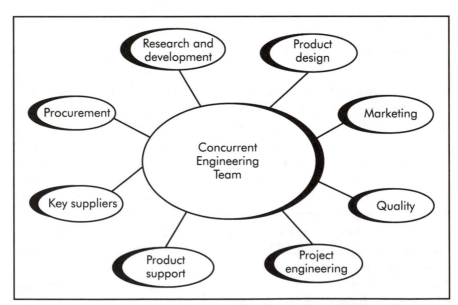

*Figure 9-7. Concurrent engineering team.*

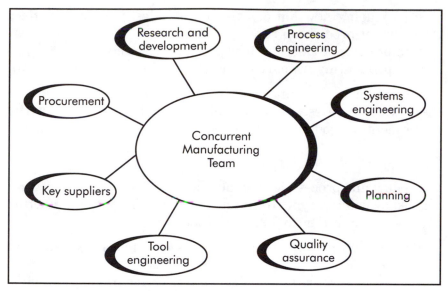

Figure 9-8. Concurrent manufacturing team.

## Technology

In either concurrent engineering or concurrent manufacturing, it is mostly technology that is targeted by multi-functional, multi-talented teams. Considering the company's interdisciplinary functions, needs, and problems, the concurrent engineering team can tackle tasks such as:

- the launching of a new product line;
- designing components into a product for better customer acceptance;
- making a product more versatile for use in other applications or industries;
- changing a product's life cycle to be more in line with the competition; and
- raising the standard of a product's quality.

The concurrent manufacturing team, on the other hand, is concerned with other issues that could or could not tie in with the concurrent engineering discussions. For example:

- changing an existing production process;
- phasing in advanced manufacturing technology;

- measuring how anticipated product-design changes affect current production runs;
- improving production-floor productivity; and
- implementing system changes (transfer versus a flex system).

The common denominator of technology is to adopt and implement what is necessary to stay ahead and be proactive for the betterment of product and process.

## Techniques

Product and process development and improvement are accomplished by using varying techniques. For concurrent manufacturing, as shown in Figure 9-9, these include:

- manufacturing agility (have processes in place to make rapid changes);
- manufacture with zero defects (strive for production without rework or scrap);
- manufacture to plant uniformity (standardize successful processes company-wide);
- manufacture to part families (produce with flexible machining systems); and
- manufacture to best practices (cost, quality, and time).

Concurrent manufacturing grows out of concurrent engineering and follows similar patterns. However, it has its identity, can function entirely through itself and, as such, will be a welcome instrument for progressive manufacturing companies.

## REFERENCE

Dundon, Elaine. 2002, June. "The Seeds of Innovation: Cultivating the Synergy that Fosters New Ideas." New York: Amacom.

## BIBLIOGRAPHY

Dubensky, Robert G. 1993. "Simultaneous Engineering of Systems." Simultaneous Engineering and Component Systems. Warrendale, PA: Society of Automotive Engineers.

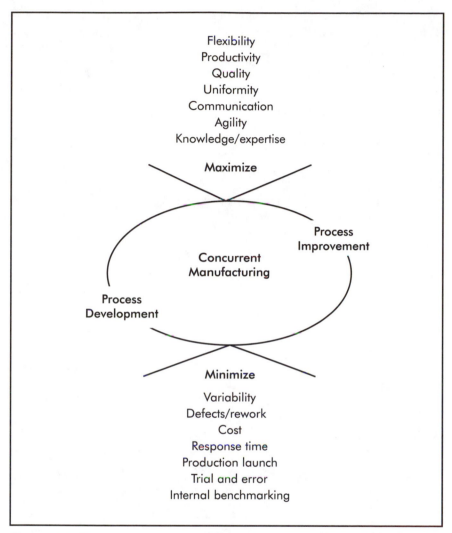

*Figure 9-9. Process improvement through concurrent manufacturing.*

Erdel, Bert. 1993. "First Part, Good Part: Zero-defect Machining." Simultaneous Engineering and Component Systems. Warrendale, PA: Society of Automotive Engineers.

Parolini, Donna L. 1995. "Global Sourcing." Global Vehicle Development Conference. Warrendale, PA: Society of Automotive Engineers.

# The Move to High-performance Machining

10

Measuring the results of machining and manufacturing determines if the results match the intended expectations. High-performance machining can be assessed by different methods and measurement techniques. The results can be qualified through multiple results and meeting specific goals. However, the common goals of every manufacturer are to keep costs down and be productive. For some, the factors of time, precision, or volume are additional considerations. High-performance machining can be perceived in different ways, qualified through different methods, and achieved by various means. This notwithstanding, there are fundamental rules, advanced technologies, and innovative ideas that can lead companies on the path to competitive, world-class machining and manufacturing.

## CASE STUDIES

The following are three examples that vividly illustrate how high-performance machining is pursued in different ways.

### Example 1: Sunglass Manufacturer

A leading sunglass manufacturer cuts lenses on special, custom-made machines, which have up to 30,000-rpm spindle speed. To prolong tool life, the fixed polycrystalline-diamond- (PCD) tipped cutting tools were redesigned. Every machining parameter was optimized with one goal in mind, the noise level at the operator's ear, which had nothing to do with the machining envelope itself. The new design cuts sunglasses now at a much lower decibel level than before. In fact, the high-pitch noise was cut in half. No major or costly machining parameter had to be changed.

Optimizing the tool's design was all it took to reduce the noise level and increase tool life simultaneously.

## Example 2: Aircraft Manufacturer

Aircraft body work is labor intensive and time consuming. Thousands of bores have to be drilled, usually with handheld drilling tools. The bores are drilled in aluminum-titanium layers with the same tool. Current machining generates impossible-to-handle chips, falling to a great extent into the hollow chamber sections of the aircraft wings and passenger compartment. Aggravating the whole situation are the coolant residues and the chip wetness. The fabricating of such complex, technologically advanced aircraft parts has to be finished without machining residues. Considering that this is done by crawling into the narrow workpiece compartments and cleaning them with rags certainly leaves much room for better and more advanced manufacturing performance.

A process was developed to prevent chips from entering the inside of the fabricated hulls during near-dry machining, dramatically increasing productivity and decreasing machining cost.

## Example 3: Engine Manufacturer

To finish-machine valve seats and guides with good tool life and achieve the desired part geometry is a challenging task for any engine manufacturer. The valve bore and guide have to be within 0.12 mil (3.0 μm) coaxially, and there is practically no room for any runout during machining. The challenges are the sintered and hardened material of the sleeve and seat ring, making tool life a constant issue. Between machine spindle and cylinder head there has to be zero centerline deviation (laser alignment). The method of pressing in the sleeves varies, and if done by force, they can end up being in a crooked position or even crack. Any centerline deviation, runout, and out-of-roundness result in premature tool failure, non-repairable workpiece damage, and frequent tool breakage.

Over the years, state-of-the-art cutting-tool technology has been changed from indexable to solid tooling, and back. Then, from PCD to cubic-boron nitride (CBN) to carbide and back again. It went from straight shanks to hollow-shank (HSK) tapers, from one-piece to two-piece designs, and to every imaginable cutting-edge geometry.

However, there was still one relevant parameter unresolved. Namely, catastrophic tool failure and breakage in case any element of the otherwise robust process goes wrong. This consideration led to a two-piece design that separates the valve guide and valve-seat tooling. Both are connected through a tool interface (a taper shank with a relatively large face area) that guarantees zero runout and repeatable accuracy. If the valve-guide tool breaks, only the front part of the tool assembly has to be replaced. The back part remains totally intact. A new valve-guide tool can be readily installed.

## SIMPLIFYING MACHINING OPERATIONS

Design engineering and production must focus on the individual technical aspects of manufacturing that determine its economical aspects. The goal is to obtain products at their least cost, while securing best quality in fit, form, function, and service. This involves simplifying and optimizing:

- workpiece configuration,
- cutting-tool material,
- machining parameters,
- realistic finish requirements,
- tool setting and inspection,
- coolant management, and
- tooling systems.

### Workpiece Configuration

Environmental, energy, and safety considerations, as well as requests from customers for more technical options have manufacturers scrambling to include more features on small parts and subassemblies. The physical properties of the workpiece material, its applicability, technical considerations, and machinability are the areas of conflicting requirements and economics. However, even the most complex parts can be productively machined with ease by selecting an advanced workpiece material, minimizing machining steps, and designing the workpiece for ease of handling, fixturing, locating, and cutting-tool access.

## Advanced Workpiece Materials

Aluminum with varying silicon content has already become the fastest growing alternative to conventional cast irons and steel. It is easy to process from the casting to the finished part and invites favorable machining techniques (high-speed and one-pass machining). Another interesting material is powdered metal. It is strong and lightweight and can be applied for near-net-shaped parts, often foregoing any additional machining. Other minor but upcoming materials are magnesium, composites, and ceramics. The key is to analyze materials that lend themselves to lowest cost production and meet design requirements.

## Minimizing Machining Steps/Facilitating Operations

The workpiece's design must secure its functionality while requiring the least amount of operations. For example, reaming can be performed instead of boring and burnishing. This preempts knowledge of the production floor's machining capabilities.

To specify a process that must be done with an outside subcontractor can be costly. It should only be done if there is no in-house resource, or if the subcontractor can produce the part less expensively. The general rule should be to design a workpiece with the machining capabilities at hand. In this case, the design staff has to seek the assistance from other manufacturing personnel.

Workpieces should be designed for ease of handling, fixturing, locating, and cutting tool access. The shorter the distance the tool has to travel from the machine, or the shorter its overhang during machining, the easier the part is to machine. The result is that the whole machine-tool/cutting-tool setup is less costly. The goal is to forego costly bushing support and feed-out arrangements or an untimely decrease in machine-tool velocity.

## Cutting-tool Material

The challenges posed by the chemical and physical composition of advanced workpiece materials have to be met by advanced cutting-tool materials (see Figure 10-1). Their selection should be based on their behavior during cutting because it is at the cutting edge where the machining process is measured. Tool life, part finish, and production time are the direct result of mating workpiece and cutting material. The metallurgical makeup of workpieces and cutting-tool materials varies greatly. The six classes have very distinguishable characteristics, as discussed in Chapter 5.

Engineers while getting bids for the most complex machine tools, often find it difficult to select the proper cutting-tool material. Lack of knowledge and information, as well as confusion related to selecting the right cutting material can result in costing manufacturers dearly. The general trend toward higher-speed machining, diverse workpiece materials, less stock removal, and better surface finishes make it necessary to carefully choose the right cutting-tool material for the right workpiece and right machining operation, as shown in Table 10-1. The selection process includes:

- choosing knowledgeable suppliers;
- having test runs performed;
- being aware that even a slight change in machining parameters has a bearing on the cutting material chosen;
- running cost comparisons of price/performance ratios rather than judging cutting materials by their initial purchase cost;
- always fine-tuning the workpiece to the cutting-tool material and machining operation; and
- applying the most advanced cutting-tool material available.

## Machining Parameters

The interdependence of machine tool, cutting tool, and workpiece material determines the machining data applied. The given cutting speeds, feed rates, and stock removal are at the core of it. Optimizing them directly affects economic and productive manufacturing.

Historically, machining in the U.S. has always been done with higher feed rates. However, now that part castings are closer to their finished shape, workpieces are easier to machine, and expected surface finishes are accomplished within fewer machining passes. There is more emphasis on speed. However, when time is

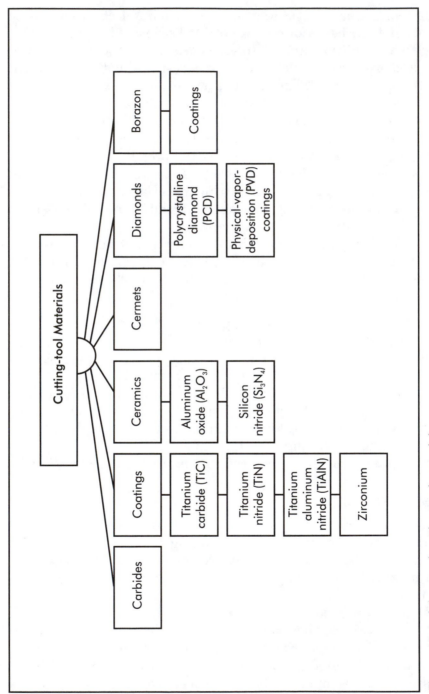

*Figure 10-1. Advanced cutting-tool materials.*

**Table 10-1. Selecting the right cutting tool based on workpiece material and process**

|  | Turning | | Boring | |
|---|---|---|---|---|
| Material | 4340 HRC 32 | | 4340 HRC 32 | |
| Cutting depth | 0.010 in. (0.25 mm) per side | | 0.03 in. (0.7 mm) per side | |
| Cutting length | 3.5 in. (89 mm) | | 4 in. (102 mm) | |
| Cutting diameter | 1.94 in. (49.2 mm) | | 1.60 in. (40.6 mm) | |
|  | Ceramic A1203 | Cermet | Carbide C6-coated | Cermet |
| Speed | 942 ft/min (287 m/min) | 942 ft/min (287 m/min) | 499 ft/min (152 m/min) | 1,201 ft/min (366 m/min) |
| Feed | 0.003 in./rev (0.08 mm/rev) | 0.003 in./rev (0.08 mm/rev) | 0.006 in./rev (0.15 mm/rev) | 0.006 in./rev (0.15 mm/rev) |
| Tool life | 15 parts | 38 parts | 2 pcs/index | 5 pcs/index |

a critical factor, feed rates have to be increased, even at the expense of the finish. In addition, feed rates are effective to control the flow of chips. At high feed rates, more pronounced material distortion and rapid deflection make the chips break easier. This is particularly important when cutting deep, blind, semi-blind, or bores to secure continuous chip discharge and forego surface scratching caused by stringy chips.

The affect that speed has during machining is temperature related. At lower speeds, the temperature at the cutting tip is below the recrystallization point, retaining the hardness of the workpiece. At higher speeds, the temperature increases, the part material softens, and cutting is more efficient because less cutting force is necessary. Higher speeds make chips flow more smoothly, thus quickly transferring the heat generated during cutting away from the cutting edge. Also, the stress on the cutting-tool material decreases, resulting in longer machine-tool/cutting-tool life. The temperature can rise excessively, causing the cutting-tool material to break down rapidly, resulting in premature tool failure.

Another influential machining parameter is stock removal. The general rule is the less stock removal, the better. However, even the finest finish cut needs some stock removal to clean the surface smoothly. Meticulous machining sometimes calls for several machining steps and that determines material removal. Particularly for operations that require special machining with special production-machine tools, such as honing or grinding, the absolute least amount of stock necessary should be left for the final operation.

The upshot is that feed, speed, and depth of cut have to be fine-tuned to one another. Controlled speeds and feeds can minimize machine-tool and cutting-tool failure (Bakerjian and Mitchell 1992).

## NEW CUTTING-TOOL TECHNOLOGY

Phasing in new cutting-tool technology can improve manufacturing productivity and quality, as well as dramatically reduce costs on the manufacturing floor. Recent technological advancements make it possible. Figure 10-2 shows some popular machining processes and the respective surface finishes they yield. It reveals some interesting overlaps. To use the chart, select the machining operation that applies the best considering the overall production flow and blueprint tolerances. For example:

- A cored workpiece bore is drilled, rough-reamed, and then roller-burnished. Substitute these processes by drilling and finish-reaming.
- A cast-iron plate is turned and finished through grinding. Substitute the production grinding operation with hard-turning (cubic-boron nitride) in a computer numerically controlled cell, which is used for regular machining anyway.
- A milling cutter over-run has to be accounted for to perform conventional slot milling. Electric discharge machining (EDM) delivers a more precise cut.
- Multiple bores are machined through single-point boring with subsequent reaming. A multi-step fine-boring tool can perform the operations in one pass.
- Reducing a blueprint finish requirement and developing a cutting-insert geometry with a primary and secondary edge might forego an otherwise expensive honing operation.

Every production floor needs to look at machining processes, compare them with the frequency of manufacturing changes, and adjust the individual operations to match cost, time, and precision requirements.

## HIGH-SPEED MACHINING

High spindle speed is a prerequisite for high-speed machining. However, it does not signify high productivity or low-cost machining. Both are a measure of fast removal rates and this means cutting speed. The user considers the cutting data primarily to minimize the cost per piece. Increased cutting speed lowers the main machining time by decreasing the cost for the machine and labor. However at the same time, the wear on the cutting blade increases; tool life goes down, which increases tool cost and noncutting time, as shown in Figure 10-3.

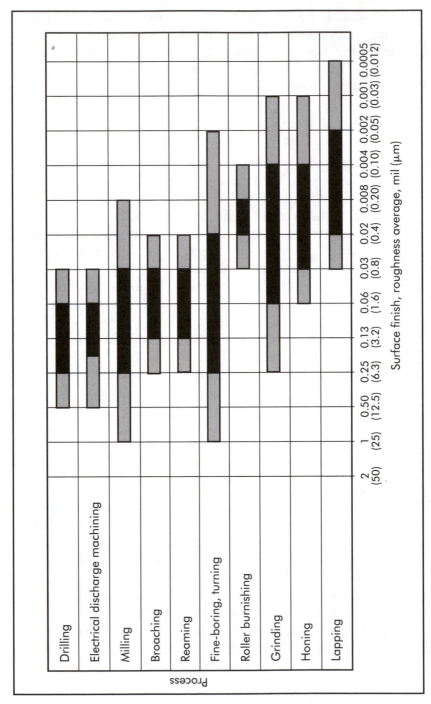

Figure 10-2. Popular machining processes and the respective surface finishes they yield.

To minimize machining cost, the optimum cutting speeds have to be determined. Much depends on the workpiece material, as shown in Figure 10-4.

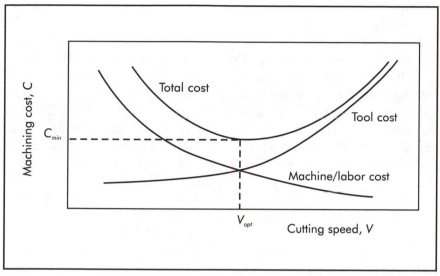

Figure 10-3. Machining cost versus cutting speed.

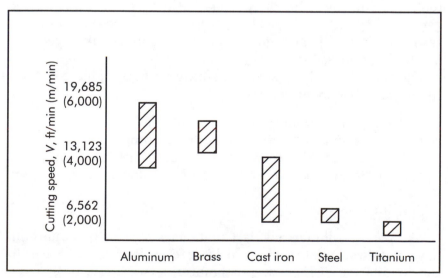

Figure 10-4. Optimum cutting speeds.

As the cutting forces diminish at the high end of the optimum cutting-speed range, the temperature increases at a greater rate while thinning the cutting chips and maximizing chip removal rates. Smaller chip loads per cutting edge result in better surface finishes. This means that roughing operations can become finish operations.

Low thermal and chemical stability disqualify carbides from most high-speed machining. Ceramics can cut up to 3,281 ft/min (1,000 m/min). However, their extreme brittleness and poor resistance to thermal and mechanical shock make their use in high-speed machining rather limited. Cubic-boron nitride (CBN) for ferrous metals and polycrystalline diamond (PCD) for nonferrous metals are the cutting materials of choice for high-speed machining, due to their superior wear resistance (high hardness, thermal, and mechanical stability). Though CBN and PCD are well suited for high-speed machining, equally important is the rigidity of the cutting tool and stability at the cutting edge.

The cutting-tool's interface with the machine tool is most critical (see Chapter 5 for more discussion on rigid tooling). Cutting-tool balance is also very important because of the relationship between force and velocity. Force is proportional to the square of rotational velocity. The centrifugal force caused by an unbalanced tool while turning at 10,000 rpm has 25 times the centrifugal force generated by the same tool at 2,000 rpm. Unbalanced forces cause severe tool chatter at high cutting speeds and make precision machining impossible.

The benefits of high-speed machining far outweigh the cost for the initial investment because:

- The metal removal rates increase five-fold.
- There is up to 70% reduction in machining times.
- There is a 25–50% decrease in the cost for machining.
- There is an exponential increase in productivity.

## Diamond Coatings

The popular nonferrous part materials, especially aluminum, have metalworking industries looking for possible alternatives to the dominating PCD cutting-tool material (see Table 10-2).

### Table 10-2. Properties of polycrystalline diamond (PCD) and chemical-vapor-deposition diamond (CVD) coatings

|  | PCD | CVD |
| --- | :---: | :---: |
| Fracture hardness | * |  |
| Hardness | * | * |
| Low coefficient of friction |  | * |
| Wear resistance (small grain size) |  | * |
| High thermal conductivity |  | * |
| Tool life (most applications) | * |  |

The chemical-vapor-deposition (CVD) diamond coating is used for certain special applications. CVD diamonds are layers of diamond crystals "grown" on silicon nitride (thin film) or tungsten carbide (thick film). While thin-film coatings seem to work well on composite material, users report better results with thick film (grain sizes higher than 2 mil [50 μm]) on aluminum.

CVD diamond's somewhat higher wear resistance, combined with its low coefficient of friction, makes it useful for turning applications. Since the smaller grain sizes are as hard as PCD's larger grain sizes, it can finish turn with excellent results. Its high thermal conductivity makes it ideal for high chip loads and high cutting speeds. CVD has been used in milling of aluminum with low silicon content (less than 8%). The problem with CVD lies in its low fracture toughness, which prohibits machining of interrupted cuts and reaming, as well as most boring operations.

CVD-diamond coatings are still in development but early reports show that they can extend tool life in some turning and milling operations. When this capability is combined with the lower purchase cost of the tool, CVD-diamond coating certainly justifies itself. If cutting-tool manufacturers can offer reliable CVD-diamond coatings at a reasonable cost, well below that of PCD, then the metalworking industry might be interested in having existing carbide tools coated. This would result in a sizable productivity boost for any manufacturing floor (Engdahl 1998).

## Tooling Interface

Manufacturing companies often overlook the importance of the machine-tool/cutting-tool interface. In particular, flexible, high-speed, and precision-finish machining need a tooling interface suited to accommodate:

- varying machining parameters;
- accuracy with frequent tool changes;
- high repeatability accuracy;
- versatility for different machining operations;
- manual and automatic handling;
- high torque transmission; and
- high rigidity (high bending strength).

The HSK taper meets these requirements, replacing less accurate interfaces, such as traditional V-tapers and variations of cylindrical shanks, as shown in Figure 10-5. As can be seen, the HSK's repeatability accuracy is outstanding. During high-speed machining, the hollow shank opens up evenly peripherally and leans itself against the mounting adapter, thus forming a very precise, stable connection. Furthermore, cutting speeds and feed rates can be increased without sacrificing quality (precision), as shown in Figure 10-6. HSK is gaining wide acceptance for almost all major applications throughout the chip-making industry because of its accuracy, higher cutting abilities, and equal or lower purchasing cost when compared to conventional interfaces.

## Hard Turning

Hard turning refers to turning heat-treated, hardened steel and perlite gray iron with a hardness of HRC 45–65. The technological conditions at which hard turning takes place are certainly different to those of conventional turning. Depth of cut of 0.008–0.012 in. (0.20–0.30 mm), feed rates of 0.006–0.008 in. (0.15–0.20 mm), and cutting speeds of 492–656 ft/min (150–200 m/min) are normal machining parameters. High cutting forces and high thermal stress are typical for the hard turning process.

Grinding was once the only operation to rough- and finish-machine extremely hard workpieces. Aluminum oxide ($Al_2O_3$), titanium carbide (TiC), and all cubic-boron nitride (CBN) grades

| Measuring points | 10 measurements, μin. (μm) | | | | | | | | | |
|---|---|---|---|---|---|---|---|---|---|---|
| | 1 | 2 | 3 | 4 | 5 | 6 | 7 | 8 | 9 | 10 |
| A | 0 | 20 (0.5) | 39 (1.0) | 20 (0.5) | 39 (1.0) | 20 (0.5) | 20 (0.5) | 20 (0.5) | 39 (1.0) | 39 (1.0) |
| B | 39 (1.0) | 39 (1.0) | 39 (1.0) | 39 (1.0) | 39 (1.0) | 39 (1.0) | 39 (1.0) | 39 (1.0) | 39 (1.0) | 39 (1.0) |
| C | 0 | 20 (0.5) | 20 (0.5) | 39 (1.0) | 39 (1.0) | 20 (0.5) | 20 (0.5) | 0 | 0 | 20 (0.5) |

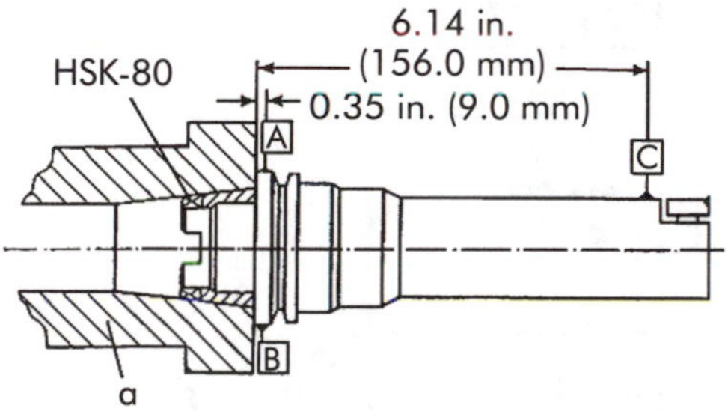

HSK-80

6.14 in.
(156.0 mm)
0.35 in. (9.0 mm)

A

C

B

a

*Figure 10-5. Repeatability accuracy of the HSK shank.*

(because of their high resistance to thermal stress and hardness) of cutting tools have begun to replace grinding in some machine shops altogether. The somewhat higher initial tool cost, compared to grinding, is more than offset by the savings in production time while achieving the same precision finishes, as shown in Figure 10-7.

What makes hard turning even more attractive for finish-machining than grinding are the inherent differences of the two processes:

- Turning is a much faster operation (shorter machining time and less part throughput time).
- Lathes offer more production flexibility.
- Roughing and finishing can be done with one clamping (CNC lathe).

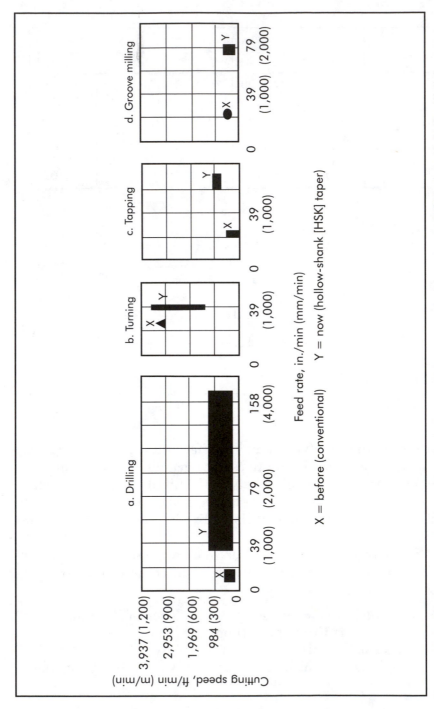

Figure 10-6. Comparison of conventional interfaces with the new hollow-shank (HSK) taper interface.

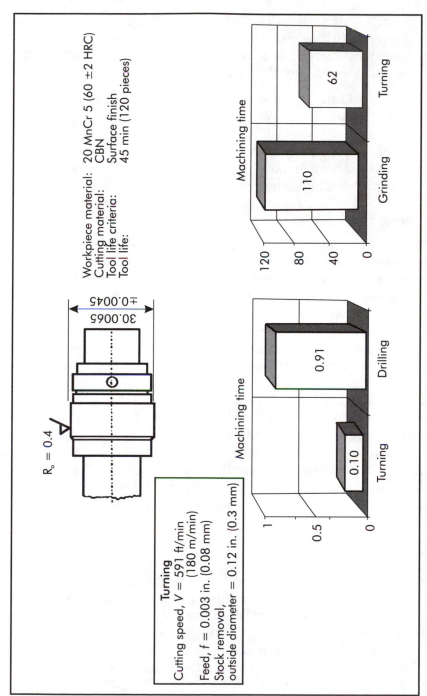

*Figure 10-7. Cost comparison of turning vs. grinding.*

- Multiple lathe operations are easier to automate through tool changes (turning cell).
- Since hard turning is done dry, there are no costs for coolant, its maintenance, or disposal.

Grinding and turning are such opposite machining operations that their substitution is not always easy. A change might be too costly to justify, especially for production floors with existing machinery. However, in cases where new machinery is required, and phasing out existing grinding machines over time is a possibility, hard turning should be the process of choice.

## Machining Passes

Progress in metallurgy, along with machine and tool design, have opened up new avenues for machining. Manufacturing needs to adopt more new machining processes because machines and tools are mostly underutilized. Many times, new machining techniques and systems are not explored. For many applications, the old-fashioned way of machining with many different operations in a step-by-step approach should be abandoned and replaced by advanced processes that use single-pass machining, as shown in Figure 10-8.

The trend to lighter workpiece materials and cleaner, more precise raw castings, combined with advanced cutting-tool technology and accurate machine tools, make heavy cuts and preparation for subsequent machining often superfluous. Lighter cuts and higher speeds encourage the reduction of machining operations. Today's tooling systems have a guaranteed competitive edge over systems of the past. They may cost more, but the cost/performance is better than before.

## Optimizing Tooling Systems

Webster defines *optimizing* as "making it as perfect, functional, and effective as possible," and this, indeed, is what high-performance machining is all about. With the knowledge outlined in the previous chapters, machining and manufacturing technologies and systems can now be combined for total optimization. To do this, all alternatives have to be considered. Advanced technologies are combined to optimize processes, as shown in Figure 10-9.

---

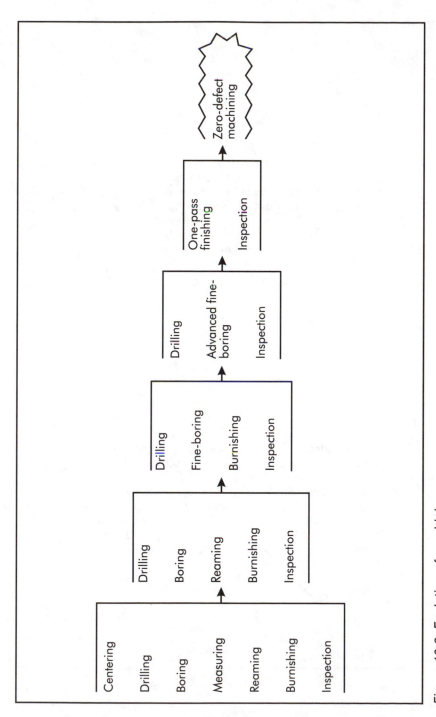

Figure 10-8. Evolution of machining.

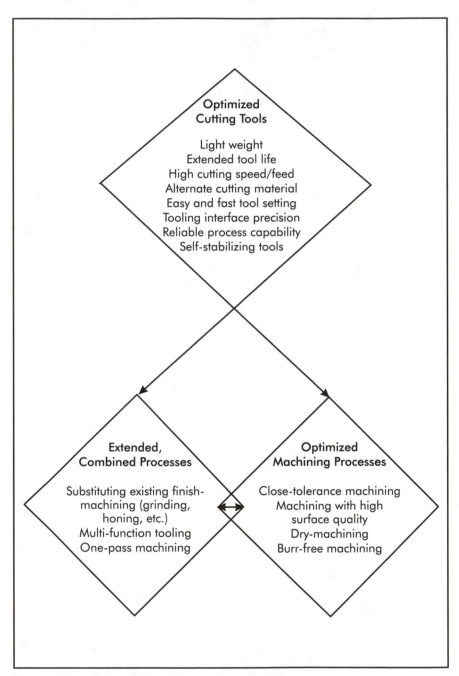

Figure 10-9. Optimized cutting tools and processes.

## First part/good part

Producing good parts right at the outset of a new manufacturing system and then continuously producing error-free, zero-defect parts is a formidable challenge. It can only be met if manufacturing is done with perpetual attention to detail and the systematic adoption of advanced technology and techniques. Most of today's value-added machining is done on extremely complex workpieces, challenging all aspects of the production floor. For example, when shaping parts through metal cutting, the attention must first be given to finish operations.

First part, good part, zero-defect manufacturing is no easy task. However, it can be done and must be pursued within certain guidelines. The allowable range of variability must be realistic.

## Design for Producibility

It is the design that determines the product's form and function. Design for producibility:

- explores simplicity;
- identifys the best production methods;
- chooses from the most suitable workpiece material;
- minimizes production;
- creates steps to minimize operations;
- plans for ease of locating and fixturing; and
- specifies the finish of the part commensurate with the process available.

## Realistic Results

What is the capability of the manufacturing floor? Can it really and realistically hold the tolerance necessary? Whether or not a part can be produced successfully is dependent on the designer's knowledge and consideration of process variation and tolerance accumulation.

Antiquated and outdated machine tools and tooling systems can not produce parts and products of a world-class standard. It is that simple. It is not enough to be quality minded, touting it, and imposing it on manufacturing. Precision manufacturing equipment is the prerequisite of producing precision parts. It is corporate management's responsibility to provide the proper tools commensurate to the quality expected of the produced parts. Only then can productive manufacturing begin and end results be cost-effectively achieved. A tight part tolerance can either be achieved with ease, predictably, and repeatedly, or with the extraordinary effort of extra processing and labor.

Statistical process control (SPC) and process capability ($C_{pk}$) recognize variations in the processes of machining and manufacturing. *Random variations*, such as temperature fluctuations and *assignable variations*, such as workpiece material and measurement variations, are the contributors to varying product quality. Random variations usually are beyond control. However, assignable variations can be targeted and dealt with. The inherent variation of testing and measurement methods and equipment is an important contributor and usually overlooked. In many cases, a part tolerance is created that cannot be measured, whether it was the equipment that was out of control or the target was totally unrealistic and unachievable.

The accumulation of tolerances within the manufacturing process is another issue to address. Again, every machine tool is built within a certain tolerance range. The same holds true for the tooling system and all other equipment contributing to the process (handling, fixturing, etc.). In addition, every machining step is done within a certain tolerance limit, as shown in Figure 10-10. The finished product incorporates the sum of all individual tolerances. For example, it is possible to consistently achieve a 0.1 mil (2.5 μm) roundness of a cast-iron connecting rod bore in high-volume production. However, this is assuming that all manufacturing modules have the inherent built-in capability for it. Manufacturing results are predictable only if tolerances are within process capability. It is prudent for corporate management to accept only the best of products. It is necessary for design personnel to specify the best possible finishes. For manufacturing staff, it is mandatory to have the capable tools on hand to produce the quality expected.

When specifying the machine system, there is a choice to be made of either a dedicated or flexible system or a combination of both. In principle, today's machine tools feature dynamic stiffness, well-balanced and stable drives, and rigid main spindles.

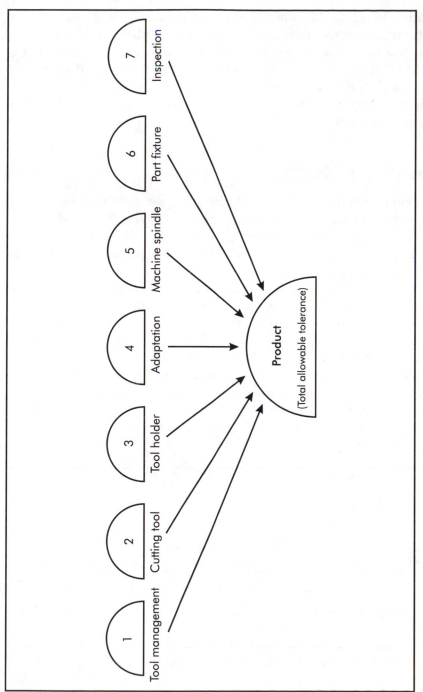

*Figure 10-10. Tolerance stack-up of process parameters.*

Taking their quality and uniformity into the design as a given, and machine capabilities as the constant within $C_{pk}$ manufacturing, the cutting-tool system is the all-important variable. It decides whether the blueprint tolerance can be cut in half to achieve $C_{pk2}$. With the high quality of today's machine tools, it is the cutting systems that make the difference in the optimization of machining processes.

## ONE-PASS MACHINING

Avoiding unnecessary roughing or post-machining operations must be the objective of every manufacturing floor. Most raw workpieces are good candidates for one-pass finishing. Advanced cutting tools can make finishing through nonproductive processing superfluous.

Investment castings are more popular because they allow the widest net-part-size range for thin walls, undercuts, and internal passages. For the forging process, near-net-shaped parts are formed through grain-flow optimization. It increases impact strength and fatigue endurance limits, which allows for a thinner wall design, as well as reduces excess weight and material stock. Overall, the trends of using more nonferrous metals and squeezing in more functions into smaller workpieces substantiate the quest for limiting machining passes (see Figure 10-11).

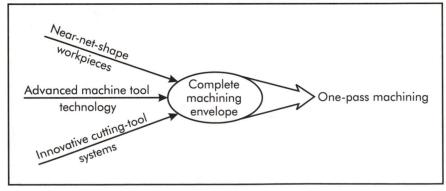

*Figure 10-11. Creation of one-pass machining.*

As cutting-tool manufacturers create the cutting-edge geometry for finish machining, other design features are geared toward total optimization of the process. Figure 10-12 shows such an approach for finish machining an engine camshaft bore. The tooling design also could be implemented for one-pass finishing of landing gears or the forks for off-highway construction equipment.

The tool in Figure 10-12 features a heavy metal tool bar for minimizing inherent cutting vibrations. It incorporates PCD inserts with primary and secondary cutting edges for super-finishes. Peripherally arranged guide pads bridge the air gaps in between the cam towers, guiding the tool rigidly throughout machining and cutting its own path. Since the guide pads are polycrystalline diamonds, they are not susceptible to chip build-up. They also minimize the amount of lubricity of the coolant. The tool's HSK design assures high repeatability accuracy, stability, and precision. The corresponding mechanical chuck provides for radial and axial adjustment to compensate for possible machine-spindle runout. This tool design and system not only eliminate pre-machining and possible post-machining, but also yield predictable machining results in terms of surface finish, bore roundness, and straightness, as well as providing a burr-free operation.

One-pass machining should be used by manufacturing because it:

- eliminates stacking up tolerances (step-by-step machining);
- decreases non-machining time (setup, tool changes);
- decreases main machining time;
- lowers tool inventory;
- promotes lean manufacturing;
- accelerates the manufacturer's response time to market shifts and changes;
- complements high-speed machining and facilitates near-dry machining; and
- provides for productivity and precision simultaneously.

Tools of the design shown in Figure 10-12 qualify for bore finish machining of any type workpiece from one side. This eliminates machine-table indexing or the provision of an opposed spindle design.

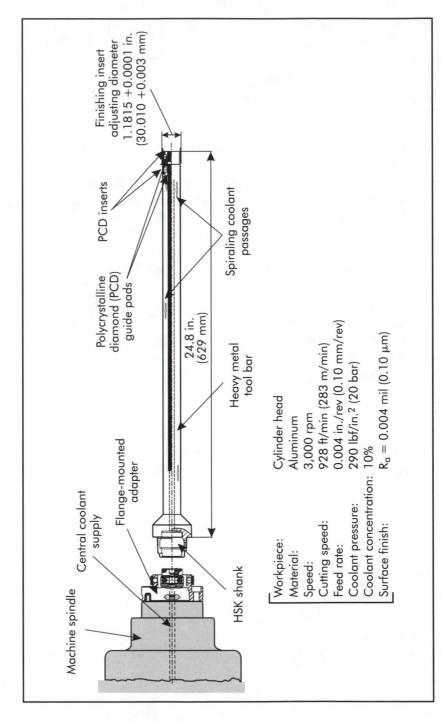

Finishing insert
adjusting diameter
1.1815 + 0.0001 in.
(30.010 + 0.003 mm)

PCD inserts

Spiraling coolant
passages

Polycrystalline
diamond (PCD)
guide pads

24.8 in.
(629 mm)

Heavy metal
tool bar

Flange-mounted
adapter

Central coolant
supply

Machine spindle

HSK shank

Workpiece:            Cylinder head
Material:             Aluminum
Speed:                3,000 rpm
Cutting speed:        928 ft/min (283 m/min)
Feed rate:            0.004 in./rev (0.10 mm/rev)
Coolant pressure:     290 lbf/in.² (20 bar)
Coolant concentration: 10%
Surface finish:       $R_a$ = 0.004 mil (0.10 μm)

*Figure 10-12. Tool design features for one-pass machining.*

Environmental pressure and the need for more productivity have manufacturing companies trying to substitute grinding operations with one-pass machining and regular production tools. For a workpiece material and

hardness exceeding 50 HRC, it is difficult. While there is expanded use of hard-turning, initial applications of hard-milling and hard-fine boring are very promising. The cutting material of choice is CBN, because of its hot hardness and resistance to abrasion.

Switching from grinding to other machining operations with geometrically defined cutting edges, such as turning, milling, and fine boring, must involve rigid machines with stiff, precise spindles, and precision cutting tools. Stock removal variations must be avoided to maintain fairly even cutting forces and control thermal stress on the tool. Contrary to popular belief, machining with higher cutting speeds in milling and turning are desirable. Cutting speeds of 1,148–1,312 ft/min (350–400 m/min) in steel and up to 3,937 ft/min (1,200 m/min) in cast iron can be applied with feed rates of about 0.005–0.006 in./rev (0.12–0.15 mm/rev). Very precise cutting-edge geometry is necessary to obtain desired surface finishes. Notch wear at the cutting edge must be minimized (low feed rates with stable, high-precision cutting edges). The surface waviness depends mostly on the machine's rigidity against vibration (see Figure 10-13).

Applications are run in labs or research and development (R&D) centers, particularly for penetrating workpieces through fine boring and reaming. This is done to empirically determine the viability of the hard-machining process (virtual manufacturing tool, collision monitoring, reliable machine control system, intricate tool design, etc.). The mold-making and automotive industries are spearheading efforts to substitute grinding and honing operations. The goals of hard machining are to use less coolant and finish machine in one pass, and to expand upon its range of applications and make full use of advanced manufacturing technology.

## COMBINATION TOOLING

Stacking up tolerances through step-by-step machining contrasts the efforts for six-sigma and $C_{pk}$ manufacturing. It prolongs

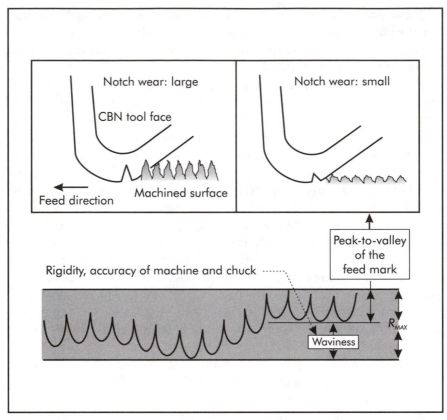

*Figure 10-13. Notch wear and surface roughness.*

the time it takes to finish the part, increases tool inventory, and leaves plenty of room for errors. Manufacturers must insist on tooling that can machine as many contours, bores, surfaces, transitions, chamfers, and other configurations of workpieces as possible with one cutting tool. This means designing one tool for several machining operations and/or using one tool to machine similar parts of the same workpiece. The approach involves high-precision combination tooling.

Of course, the purchase prices for combination tooling are higher per piece than for single-purpose tooling. Purchasing has to evaluate the initial investment on the ratio of price/performance. It does not take much convincing on the parts of engineering and manufacturing to point out the striking benefits of such tooling systems.

Given advanced technology, even the most complex machining configurations can be designed into the tool. Figure 10-14 illustrates the finish machining of transmission cases with a multi-step tool. It features a number of technologically advanced tooling characteristics.

- The precision-ground inserts are fine-tuned to one another for exact location.
- An aluminum or magnesium tool body reduces weight and is surface-hardened for abrasion resistance.
- The two-piece design has guaranteed zero runout and repeatability accuracy through a precise interface. The tool is stabilized through the front part (more wear). It is more economical and user-friendly to just replace the front part in case of a crash due to machine misalignment or for service.
- The HSK taper assures repeatability, accuracy, and accurate transmission of torque and higher cutting speed. Defined and guided coolant passages are provided to all cutting edges individually to aid in achieving the desired surface finishes and secure proper chip discharge.
- Peripherally arranged guide pads separate the cutting and guiding of the tool. They are of PCD material to minimize wear and allow for use of coolant with minimum lubricity.

Sometimes parts feature outer and inner diameters that have to be absolutely concentric to each other. An example would be crankshaft face bearing bores (as shown in Figure 10-15) or transmission valve-body sleeves. Only a multi-step combination tool can guarantee a successful fine-boring operation.

Going into the solid and finishing the workpiece with one tool in one pass is something manufacturing and process engineers dream about. Whether the workpieces are of ferrous or nonferrous metal, tools can be developed that incorporate drilling and reaming or fine-boring for a robust process and precision part finishes. Such a tool design would feature a self-centering spade-type drill for the front part. Trailing it would be an indexable insert supported by guide pads. The tool body would end in an HSK mechanical taper.

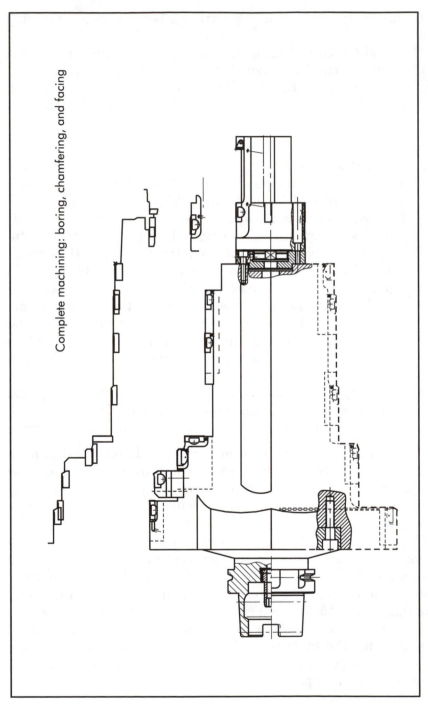

Complete machining: boring, chamfering, and facing

*Figure 10-14. Multi-step tool.*

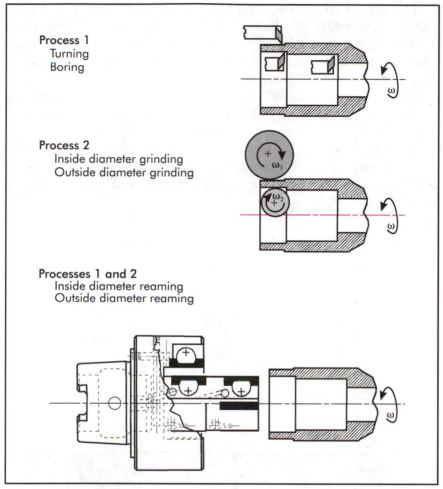

**Process 1**
Turning
Boring

**Process 2**
Inside diameter grinding
Outside diameter grinding

**Processes 1 and 2**
Inside diameter reaming
Outside diameter reaming

*Figure 10-15. Outside and inside-diameter, one-pass machining.*

# INTERPOLATION TOOLS

Combining the superior traits of machine tools and cutting tools in terms of dynamics, accuracy, and power invites machine processes involving circular (helical) milling and interpolation. In circular milling, bore diameters are incrementally generated. Interpolation takes circular milling one step further by adding another dimension while describing more geometric forms to otherwise helical, cylindrical, and radius shapes.

The idea of circular milling is to use one tool to finish several other bores of different diameters. Preferably, the workpiece stays in one clamping position, the spindle is loaded with one tool, and the machine axes moves to different parts of the workpiece for finish machining.

Helical and circular milling, in particular with large bores and/ or heavy stock removal, offers reductions in cycle time and tooling costs. Circular milling tools can be used to perform tasks through interpolation. For example, where one tool, through helical interpolation, counterbores and thread mills. It can do so over a wide range of diameters with the same thread pitch, as shown in Figure 10-16.

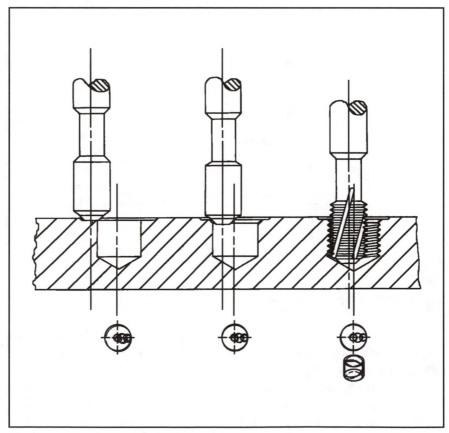

*Figure 10-16. Thread milling through interpolation.*

## ONE-STEP THREAD MILLING

One-step thread milling reduces the number of tools needed, limits non-machining time, and eliminates tool changes, as shown in Figure 10-17.

New tooling developments incorporate drilling for finishing threads by going into the solid. Circular interpolation can take on different machining tasks. Besides boring and threading, it can do slotting, undercutting, and surface milling. Figure 10-18 shows a grooving tool that is held in a hydraulic chuck and finish-machined through circular interpolation.

Circular milling cutters and interpolation tools are breaking new ground in the way the outer and inner contours of precision workpieces are finished. However, there are a few important considerations to take into account to successfully apply these tools:

- The machines have to be very rigid and dynamically stiff.
- The machine spindle and the tooling assembly have to be balanced at grade G2.5.
- The path the tool describes is directly related to the precision of the machine drives.
- When machining bores and removing stock of more than 0.098 in. (2.50 mm) on the side, there is the chance of recutting chips. High-pressure coolant or air must be provided to remove the chips.
- The tool must have a certain inside diameter/outside diameter (ID/OD) workpiece engagement. High engagement decreases inherent vibration but also increases the thrust force applied.
- For interpolation at extremely high speeds above 12,000 rpm, hollow milling tools might be considered. A reduced mass means reduced centrifugal force, which then reduces the radial load on the spindle.

## HYBRID PROCESSES

Preparing the manufacturing floor for the use of so-called "super-processes" and implementing them competently leads to technological and market leadership. A *super-process* is defined as using the ultimate level of technology known and:

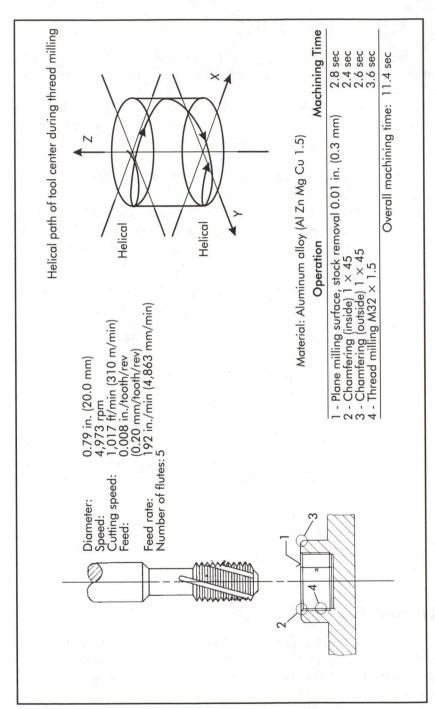

Figure 10-17. One-step thread milling.

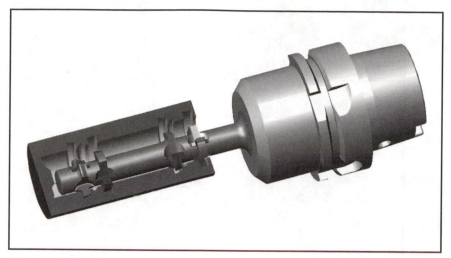

*Figure 10-18. Circular milling tool.*

- yielding high productivity, high precision, and high reliability;
- achieving results economically, that is, with a favorable price/performance ratio;
- infusing new innovation as it becomes available; and
- maintaining the process at the technological edge.

After making the decision to phase-in super-processes on the production floor, a long-term commitment is necessary. Only continuous involvement can secure leadership in technology and, with it, market leadership (see Figure 10-19). To keep up with super-processes requires knowledge. This not only includes knowledge of the processes used in-house but also of competitive processes. It includes peripheral knowledge not directly related to the current process but to support activities such as research and development, process simulation, rapid prototyping, and information technology.

New technology matures quickly because of rapid innovation. This makes it extremely difficult for one manufacturer to create new technology for existing processes. Extensive research on promising developments is essential and should be done with other partners either from academia, subcontractors, or in conjunction with government initiatives, which usually pool technology leaders for a new project.

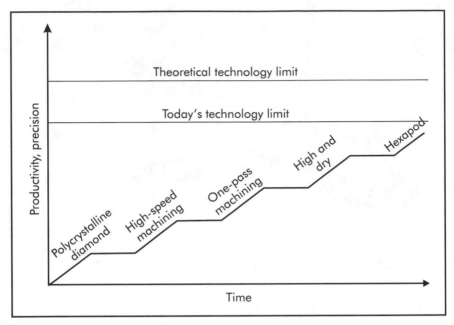

*Figure 10-19. Super-machining processes.*

Satisfactory use of one process leads to developing new technology, which in turn, stimulates new innovative processes. However, while product follows process, it is the demand of the product that initiates the development of new processes. Product and process lead and follow alternately. Hybrid processes make the manufacturing process more economical for:

- difficult-to-machine material such as titanium alloys; or
- reaction-bonded silicon nitride, used primarily in aircraft; or
- the aerospace industry for light weight and heat resistance; or
- ceramics used for roller bearings and cutting material, micro-turbines, or other industrial applications.

Products that can sustain extreme loads and application conditions such as abrasion, heat, friction, and mechanical stress have to be subjected to extreme machining conditions or nonstandard machining parameters, which involve different physical and chemical reactions during the process. Preferably, the process is productively and economically done with regular production machines and production tooling.

Difficult-to-machine materials are traditionally machined through grinding, lapping, or honing. These processes are performed with non-geometrically defined cutting tools or mandate the use of cutting fluids and involve extraordinarily long machining times.

Hybrid processes use a second form of energy that assists the original form, which is usually mechanical. The goal is to improve over an existing traditional robust process by adding another robust process. The combination of the two processes overcomes the original technical limitation. The physical and chemical agents to consider are:

- laser,
- ultrasound,
- liquid nitrogen,
- electrochemicals,
- waterjets, and
- microwaves.

## Laser-assisted Turning

The challenge of machining extremely hard materials on regular production runs has led to the development of a method that incorporates a laser beam as an add-on package to existing machines.

Silicon nitride ($Si_3N_4$), with its extremely high tensile strength, makes any cutting material break down quickly due to thermal stress. However, by using laser-supported cutting, the cutting area can be pre-warmed. The resulting increase in temperature transforms the physical-mechanical characteristics of the workpiece material by making it more pliable. This decreases its tensile strength and improves machinability.

Laser-assisted machining is a hot machining process (1,472–5,432° F [800–3,000° C]) that not only lowers the cutting forces but also reduces tool wear, decreases vibration, and allows for higher removal rates (see Figure 10-20).

Using laser assistance, turning operations exhibit a 50% reduction in cutting forces due to increased temperatures in the chip cross-section, which is achieved by softening the workpiece material. The wear on PCD and CBN cutting-material ceramics

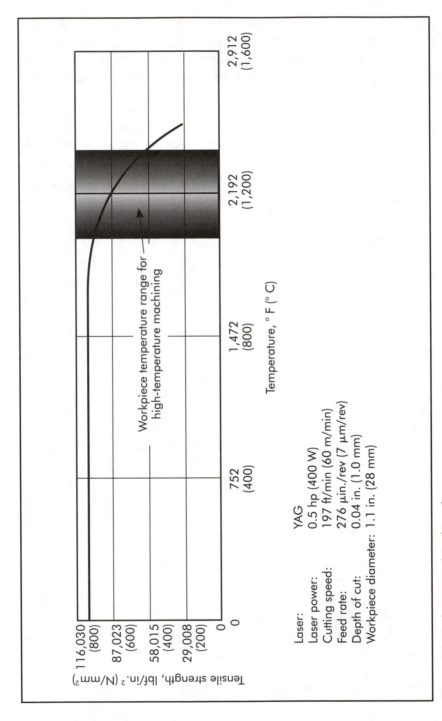

Figure 10-20. Laser-assisted turning.

is somewhat higher when turning, but the surface finishes are comparable to those of grinding operations.

Generally, machinability can be improved with the aid of laser treating. Tests of milling ceramics with laser assistance reveal that cutting forces can be reduced by up to 70% while tool wear decreases by about 80%. Milling of steel results in cutting-force reductions of 30–70% and substantial tool life increases when compared to conventional milling operations.

## Cryogenically Cooled Turning

Titanium-aluminum-vanadium (Ti-6A-4V) and reaction-bonded silicon nitride are representative of extremely difficult-to-machine materials. Both are used in the aerospace industry due to their hardness, high strength, low thermal expansion, and heat-treating ability. However, they both exhibit low thermal conductivity. The high temperature generated during turning causes the cutting material to break down, leading to premature tool wear. Low metal-removal rates and the cost for the tools have led to a different method of machining. This system, as illustrated in Figure 10-21, uses a cryogenic cooling system that induces liquid nitrogen. This is done near the rear end of the cutting insert through a cavity mounted on the toolholder and forms a chamber through which the liquid nitrogen flows in and out constantly.

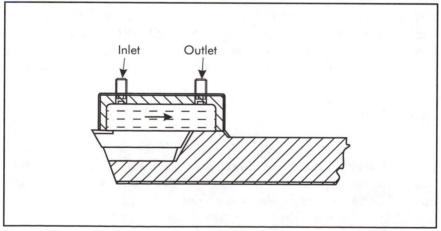

*Figure 10-21. Principle of cryogenically cooled turning.*

Turning reaction-bonded silicon nitride with CBN and cryogenic coolant lowers tool wear substantially and improves surface finishes, since the cutting edge does not break down as quickly during the first cuts. Dry machining of Ti-6-A-4V with a regular carbide insert and liquid nitrogen shows that the insert's flank wear occurs at a much slower pace.

Machining of difficult materials becomes more manageable and more productive when performed dry with the aid of a liquid nitrogen agent. The cooling effect reduces tool wear and prevents softening of the cutting insert.

## Sound-wave-assisted Grinding

Ceramics are characterized by their hardness, tensile strength, relative low density, and high thermal and overall chemical stability. These characteristics cause them to be a great challenge for contemporary finish-grinding operations. Necessary high cutting forces yield unsatisfactory surface finishes due to material damage. To avoid catastrophic workpiece damage, the only viable machining alternative is lapping. However, this process is usually unacceptable because of its low economics and productivity.

One process induces high-frequency oscillating movements superimposed during grinding. The effect is a reduction of the cutting forces, which in turn allows increasing the feed rate. Minimizing the contact between the workpiece and grinding wheel prolongs tool life and dramatically lowers the thermal stress on the workpiece. The oscillating waves also provide for automatic coolant control, as shown in Figure 10-22. Producing precision carbide inserts through the application of sound waves will be one of the technology breakthroughs in cutting tool technology.

# CONCLUSION

The complexity of parts and end products, combined with the physical and chemical demands put on them, pushes advanced machining processes to their limits. The combination of mechanical and other energies opens new paths for manufacturing as a whole. Intertwining heretofore unrelated sources of chemical and physical origin will set completely new manufacturing standards.

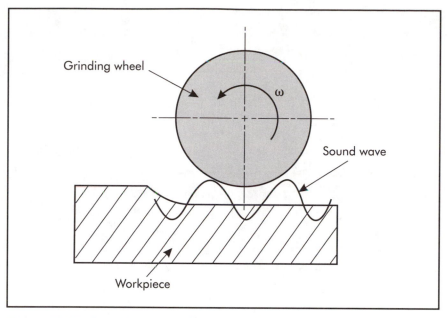

*Figure 10-22. Sound-wave-assisted grinding.*

Future machining technology will still involve machines and tools. Today, new super-processes are only applied sporadically. Tomorrow they will dominate our manufacturing landscape (WZL-Aachen 1999).

## REFERENCES

Bakerjian, Ramon and Mitchell, Philip, eds. 1992. Volume 6, *Design for Manufacturability. Tool and Manufacturing Engineers Handbook*, 4th Edition. Dearborn, MI: Society of Manufacturing Engineers.

Engdahl, Chris. 1998. "CVD Diamond-coated Rotating Tools for Machining Advanced Materials." Dearborn, MI: National Center for Manufacturing Sciences (NCMS) Workshop Series.

WZL-Aachen. 1999, June. *Production Technology*. Fraunhofer Institute. Aachen, Germany: Shaker Publication Company.

# Practical Applications

**11**

This chapter will discuss some major machining applications and realistic production runs. Advanced engineering and manufacturing technologies are converted into real production and process improvements.

## GUIDED FINE-BORING

Fine-boring tools designed with guide pads and indexable inserts have been very successful in finish machining bores. The reason for their overwhelming success is the separation of cutting and guiding elements, yielding superior surface finish and close tolerances, as shown in Figure 11-1. The performance level of these tools is further enhanced through the use of cutting materials, such as cubic-boron nitride (CBN) and polycrystalline diamond (PCD), resulting in increased cutting-speed capability. The hollow-shank (HSK) taper tool interface provides the minimum tool runout and high repeatability accuracy. Increased cutting speeds are made possible by using advanced guide-pad material, such as PCD, which reduces material build up and wear. The achieved cutting speeds in cast iron range from 984 ft/min (300 m/min) with carbide inserts up to 2,953 ft/min (900 m/min) with CBN. For steel, cutting speeds are up to 1,148 ft/min (350 m/min). They are almost unlimited for machining aluminum (the spindle rpm capability of the machine tool is the limiting factor).

A further performance improvement is achieved by adding a second insert to the design, thus allowing for increased feed rates in addition to high cutting speeds. The most promising results are achieved with the micro-machining blade arrangement. In this

*Figure 11-1. Tool head.*

case, both blades are set to slightly different radial and axial dimensions, thus leaving just a few hundredths of a millimeter stock for the finishing insert. The high cutting speed is combined with twice the feed rates of the traditional single-blade design (see Figure 11-2). At the same time, good surface finishes and excellent tool life are achieved.

Since one insert performs roughing while the second removes only minimum stock for finishing, a stable process is achieved. Applying the same cutting speed, the machining time is less than half due to the increased speed. Quality is improved, too. For example, successful applications on connecting rods or wheel-brake cylinders show that any post-machining can be eliminated.

*Figure 11-2. Finish machining wheel-brake-cylinder bore.* (Courtesy MAPAL, Inc.)

## PRECISION HARD MACHINING

Excellent tool performance occurs when single- and twin-bladed tools with PCD guide pads are used to machine hardened steel (up to 62 Rockwell hardness [HRC]) with CBN as the cutting-tool material. Cost savings are accomplished through elimination of the traditional final grinding or honing operation. One example is the finish machining of pinion gears where reaming is compared to grinding or single-point boring. Reaming is the more economical process due to the provision of PCD guide pads. Another example is machining gear blanks of case-hardened steel with 60 HRC. Table 11-1 shows the tool data for the design with hexagonal CBN inserts and PCD guide pads. The high-performance characteristics result in tool-cost savings and increased cycle time with the use of regular production machines.

### Table 11-1. Applied machining parameters

| | |
|---|---|
| Machine: | Computer-numerically controlled (CNC) lathe |
| Coolant: | 10% emulsion |
| Cutting speed: | 656 ft/min (200 mm/min) |
| Speed: | 1,300 rpm |
| Machining results: | Surface finish = Rz = 0.04–0.08 mil (1.0–2.0 μm) |
| Bore straightness: | 0.19 mil (3.0 μm) |
| Tool life: | 700 bores |
| Cost reduction: | 50% more than conventional machining |

## MACHINING COMPACTED GRAPHITE IRON

Compacted graphite iron (CGI) is a material of interest to engine manufacturers. It is used for engine parts, especially diesel engine blocks and heads, because of its superior physical and mechanical characteristics. CGI's tensile strength is about 75% higher than that of gray cast iron and it has double the fatigue strength. This means less workpiece material is necessary, making engines lighter and more comparable to aluminum castings. Moreover, CGI's molecular makeup provides excellent thermal conductivity and vibration-dampening properties. However, machining this compact material can be difficult.

Extensive research and test machining has revealed that high cutting speed in conjunction with CBN cutting tools on CGI can not be economically applied. To meet high cycle times, feed rates have to be elevated. One of the most difficult engine applications is machining cylinder bores, especially when considering realistic tool-life expectations, desired surface finishes, and cost minimization. Excellent results can be achieved with the tool design shown in Figure 11-3. It features six C4 hexagonal, coated carbide inserts with four for semi-finishing. Polycrystalline-diamond guide pads secure proper guidance of the tool during cutting. The micro-machining design of the tool combines high feed rates with good surface finishes, eliminating post-machining processes, as shown in Figures 11-4 and 11-5.

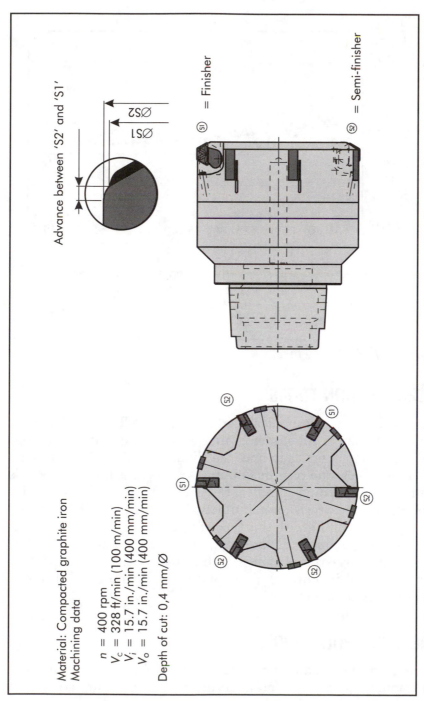

Material: Compacted graphite iron
Machining data

$n$ = 400 rpm
$V_c$ = 328 ft/min (100 m/min)
$V_i$ = 15.7 in./min (400 mm/min)
$V_o$ = 15.7 in./min (400 mm/min)

Depth of cut: 0,4 mm/Ø

Advance between 'S2' and 'S1'

⌀S1
⌀S2

⑤ = Finisher

⑤② = Semi-finisher

*Figure 11-3. Tool design for machining compacted graphite iron. (Courtesy MAPAL, Inc.)*

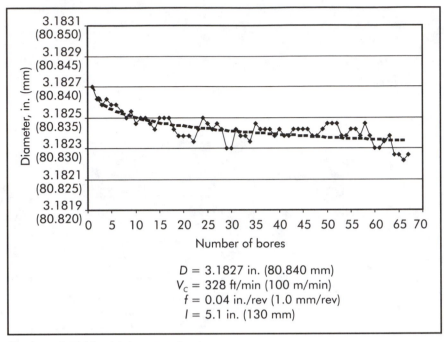

*Figure 11-4. Machining results.*

## INTERPOLATION TOOLS

Interpolation tools are designed with fixed PCD or CBN as combination tools for finish-machining bores, bore transitions, flat surfaces, and chamfers in one (see Figures 11-6, 11-7, and 11-8). With such tools, one-pass machining is made possible with today's high-quality machine spindles, the accuracy of the HSK tool interface, and the precision manufacture of PCD and CBN tooling. One-pass machining with this type of tooling reduces non-machining time (setup and tool changes), eliminates stacking up tolerances (step-by-step cutting), promotes lean manufacturing (lowering tool inventory), and complements the capabilities of high-quality machine tools.

## PRECISION ROUGHING

Intertwining production facilities that manufacture identical and similar parts in different locations force manufacturers to

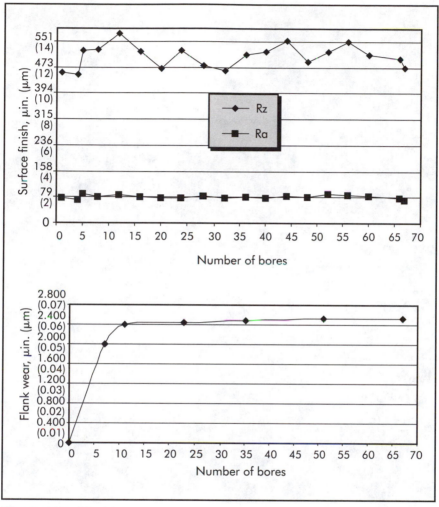

Figure 11-5. Machining results.

move toward commonality of parts, processes, and tooling. The specification of International Organization for Standardization (ISO) inserts has long been part of such efforts. Typically, ISO-type tooling is used for roughing operations and machining parts with relatively wide-open tolerances.

In cases where ISO tools rough or pre-machine a part, quality for roughing is often overlooked. However, it is as important for the first machining pass as it is for the final pass itself, because

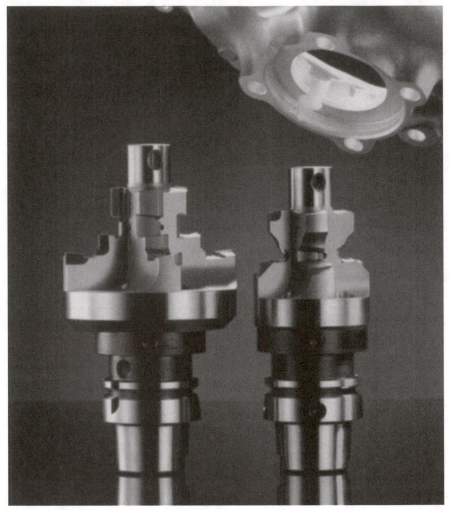

*Figure 11-6. Four-bladed PCD tool.* (Courtesy MAPAL, Inc.)

the roughing process affects the finishing result. Furthermore, there is often a poorly defined definition for what constitutes roughing and finishing operations. This is because traditional roughing might not be good enough and another finishing pass is needed. Therefore, it makes sense to develop a family of tools that combine cost effectiveness with flexibility and precision.

Quality can only be achieved if ISO-type tooling uses precision-milled pockets for insert placement, precision-ground inserts, and

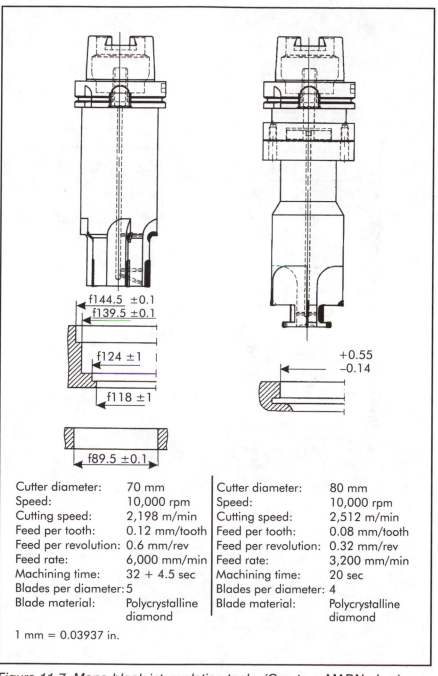

| Cutter diameter: | 70 mm | Cutter diameter: | 80 mm |
|---|---|---|---|
| Speed: | 10,000 rpm | Speed: | 10,000 rpm |
| Cutting speed: | 2,198 m/min | Cutting speed: | 2,512 m/min |
| Feed per tooth: | 0.12 mm/tooth | Feed per tooth: | 0.08 mm/tooth |
| Feed per revolution: | 0.6 mm/rev | Feed per revolution: | 0.32 mm/rev |
| Feed rate: | 6,000 mm/min | Feed rate: | 3,200 mm/min |
| Machining time: | 32 + 4.5 sec | Machining time: | 20 sec |
| Blades per diameter: | 5 | Blades per diameter: | 4 |
| Blade material: | Polycrystalline diamond | Blade material: | Polycrystalline diamond |

1 mm = 0.03937 in.

*Figure 11-7. Mono-block interpolation tools.* (Courtesy MAPAL, Inc.)

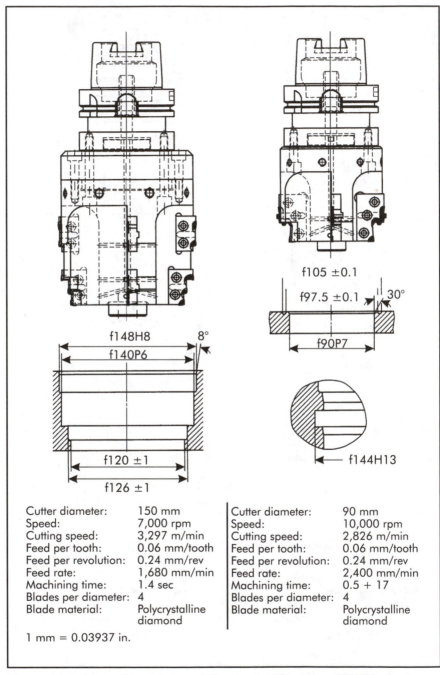

f105 ±0.1

f97.5 ±0.1 ⟍ 30°

f90P7

f148H8    8°

f140P6

f120 ±1

f126 ±1

f144H13

| Cutter diameter: | 150 mm | Cutter diameter: | 90 mm |
|---|---|---|---|
| Speed: | 7,000 rpm | Speed: | 10,000 rpm |
| Cutting speed: | 3,297 m/min | Cutting speed: | 2,826 m/min |
| Feed per tooth: | 0.06 mm/tooth | Feed per tooth: | 0.06 mm/tooth |
| Feed per revolution: | 0.24 mm/rev | Feed per revolution: | 0.24 mm/rev |
| Feed rate: | 1,680 mm/min | Feed rate: | 2,400 mm/min |
| Machining time: | 1.4 sec | Machining time: | 0.5 + 17 |
| Blades per diameter: | 4 | Blades per diameter: | 4 |
| Blade material: | Polycrystalline diamond | Blade material: | Polycrystalline diamond |

1 mm = 0.03937 in.

*Figure 11-8. Mono-block interpolation tools.* (Courtesy MAPAL, Inc.)

insert adjustability. This is especially true for performing a one-pass, rough-machining operation with complicated, stacked-up bore configurations. For the multi-step tools, the relative position of the inserts to one another and adjustability within 0.0020–0.0039 in. (0.050–0.100 mm) are important criteria. These tool characteristics guarantee long tool life, precision machining, and tolerance flexibility.

## ADVANCED MILLING OPERATIONS

Hard milling of homokinetic velocity joints is an improvement when compared to other more traditional machining processes. In the past, velocity joints were sequentially machined through boring and milling the soft slate, then hardening, and finally grinding the racetracks. This is time consuming and costly. It can be substituted by hard milling with a precision ball-nose cutter (see Figure 11-9), which is a simplified and cost-effective process. Today, these homokinetic joints are precision forged, hardened, and then machined in the finish-hard state with one pass.

The use of CBN inserts, as shown in Figure 11-10, in conjunction with advanced flexible computer numerical control (CNC) machining centers make it possible to not only forego costly and time-consuming grinding operations, but to describe complicated machining contours (elliptic, gothic, etc.). Cost savings of up to 60% have been reported.

For face milling, the measure of performance is:

- high cutting speed,
- high chip loads, and
- complete chip discharge.

The measure of precision is:

- surface flatness,
- burr-free surface, and
- predictable surface finish.

To accommodate machining nonferrous metals, advanced milling cutters feature aluminum bodies with built-in cartridges, which are PCD tipped with high-precision-ground cutting geometry. To do one-pass machining, the cutters are stacked alternately for

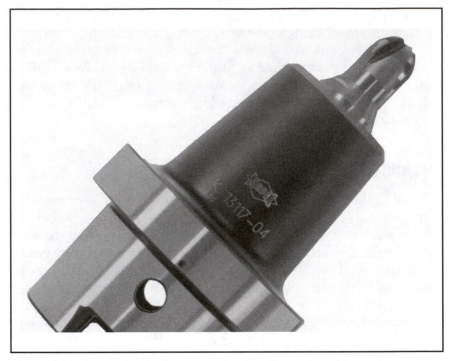

*Figure 11-9. Precision ball-nose cutter.* (Courtesy MAPAL, Inc.)

roughing and finishing. The milling tool is held with an HSK arbor and precision balanced.

Maximum permissible speeds for given cutter diameters were empirically determined through tests to assure safe use on the production floor. Figure 11-11 shows the permissible parameters. The reason for the steep incline of the cutting-speed curve is that within a diameter range of 3.2–6.3 in. (80–160 mm), the determining factor is the mounting of the cutting inserts. However, above the 6.3-in. (160-mm) diameter, the tool body is the critical element since it would break first.

Steel deflectors, built into the aluminum body and located right behind the cutting inserts prevent tool-body erosion due to chip abrasion. Optional built-in brushes are activated by air or coolant pressure to create a burr-free surface finish. This feature eliminates the need for an extra tool and/or the machine station for that operation.

High-speed Machining

*Figure 11-10. Ball-end cutter with CBN insert.*

For chip storage and a chip-free machining environment, the milling cutters can be housed in a steel enclosure (see Figure 11-12). The chips' natural dynamics of force and flow discharge them out of the machining area.

## ACTUATING TOOL SYSTEMS

On dedicated transfer lines, conventional feed-out tools have long been the answer for difficult-to-access workpiece configurations. However, with the manufacturing floor now dominated by CNC machining centers, new tooling systems have made inroads, addressing the need for tool and part contact through life bearings, coolant pressure, airflow, or centrifugal force.

A high-performance, twin-tracing tool allows machining of aluminum transmission housings with different internal profiles by reprogramming the axis movement without changing the tool. This design eliminates the problem of distorting the transmission housing if a multi-cutting-edge boring bar is used.

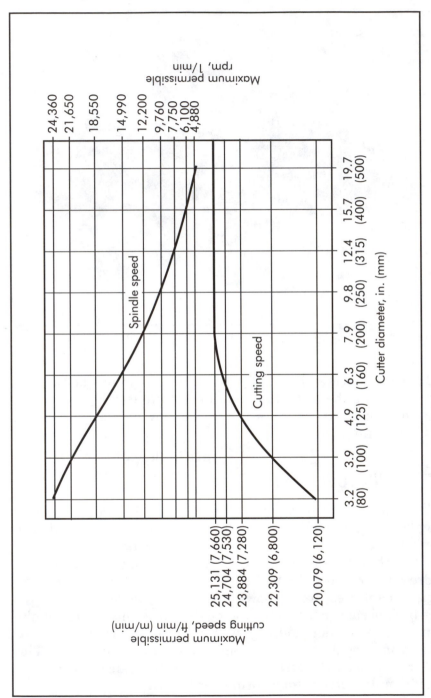

*Figure 11-11. Maximum permissible cutting speeds.*

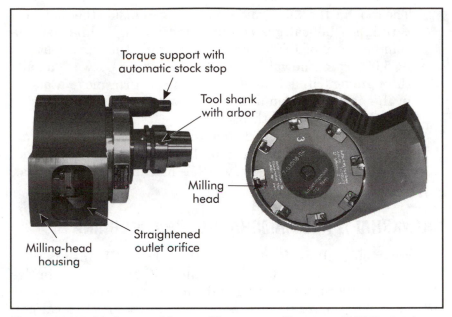

*Figure 11-12. High-speed face milling cutter. (Courtesy MAPAL, Inc.)*

The goal of achieving a torsion-free machining operation without chip problems has been fully reached. Two PCD-diamond cutting edges work simultaneously at a cutting speed of 3,280 ft/min (1,000 m/min). It has a feed rate of 1.200 mil (0.03-mm)/cutting edge and cutting depths of 0.04–0.12 in. (1.0–3.0 mm) on the radius. The tool is dynamically stable and holds a tolerance of 0.0010 in. (0.025 mm) on the radius over the full profile of the internal diameter. The copying range for this tool is 5.9–8.7 in. (150–220 mm) in diameter.

Precision machining of valve seat/guides for engine cylinder heads is a critical operation. Bores are either generated by a process such as turning, or plunged by fine-boring and reaming. The critical finishes are bore roundness and concentricity between seat and guide. Other criteria, because of the overwhelming use of flexlines instead of dedicated transfer-line stations, are tool repeatability and accuracy. Unlike transfer lines on most machining centers, an additional axial feed for the valve-guide tool is not available. A self-contained tool utilizes its natural kinetic energy and functions as follows.

- The tool for the valve guide is positioned inside the tool body through a helical gas compression spring, while using a spindle speed of 1,000 rpm for machining the valve seat.
- At 5,000 rpm, the valve-guide tool is moved outward axially through centrifugal weights, which, in turn, move a piston in the axial direction at a precisely defined feed rate using the oil and a coil as a throttle.
- Retracting the valve-guide tool to a speed of 1,000 rpm is accomplished in less than a second to position the entire feed-out tool for the next stroke, which is machining the valve seat.

## CRANKSHAFT- AND CAMSHAFT-BORE MACHINING

A two-step tool, made of heavy metal, has PCD inserts with primary and secondary cutting edges and PCD guide pads to bridge the air gap between the journals and guide the tool rigidly through the full machining pass. The tool is held in the mechanical hollow-shank (HSK) taper and the corresponding radial adjustment adapter. Use of the two-step tool (see Figure 11-13) eliminates any post machining, and yields precise and predictable results in surface finish and bore geometry.

Crankshaft and camshaft bores typically were machined with conventional line boring bars. A more accurate, user-friendly process is secured with an advanced tooling system featuring peripherally arranged PCD guide pads and four carbide inserts. The two-piece design is made possible with the precision HSK connection, as shown in Figure 11-14. Radial adjustable adapters compensate for any spindle runout. The advantages of this system compared to a line boring bar are:

- reduced downtime due to short tool-change time;
- better size control in all journals;
- no need for readjustment over entire tool life; and
- more user-friendly.

One measure for camshaft and crankshaft bore finishes is the centerline deviation measured from the first to last journal. Here, too this advanced process proves to be clearly superior.

*Figure 11-13. Crankshaft-bore tool design.* (Courtesy MAPAL, Inc.)

## HELICALLY FLUTED MILLING

Finishing axle-pivot arms with ISO standard-pocketed tooling in the IT9 (international tolerance band) range with predictable bore geometry is only possible with ISO inserts that can be evenly adjusted radially. This ensures that all inserts cut at the same time. With tangential inserts, the tools build sturdier, chip flutes can be better optimized, less cutting force is required, and cutting vibration is minimized.

All truck and car parts for the main axle are made of hard cast iron. High production rates, relatively low insert tool life, and re-laxed finish tolerances invite the use of precision ISO-standard, pocketed adjustable tooling to safely and predictably improve these conditions.

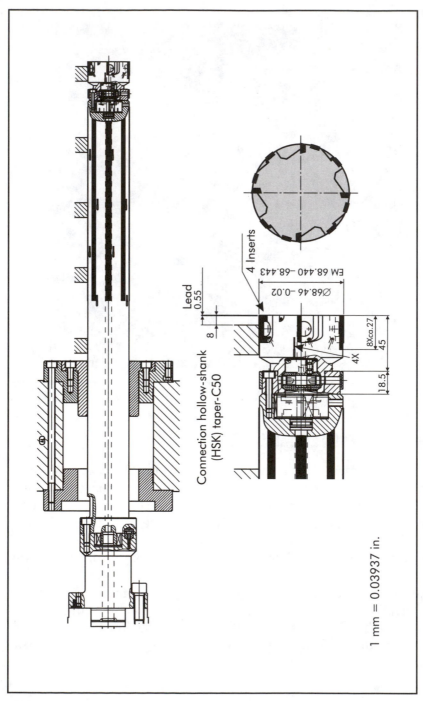

Connection hollow-shank
(HSK) taper-C50

Lead
0.55

4 Inserts

EM 68.440–68.443

Ø68.46 –0.02

8Xca.27

45

4X

18.5

8

1 mm = 0.03937 in.

*Figure 11-14. Precision HSK connection. (Courtesy MAPAL, Inc.)*

The aircraft industry is one of the premier customers for milling operations. Often, the given machining parameters are extremely demanding. Only helically shaped milling cutters can accommodate the requirements shown in Figure 11-15.

Tool optimization includes: the use of insert spacers to form a helical shape, setting the helix's angle, cutting-edge-geometry preparation, and precise manufacture. The result is a process with high chip removal rates and very fast finish machining, while guaranteeing smooth cutting at reduced noise levels.

## MACHINING WITH MULTI-CUT TOOLS

The process of precision machining hard workpiece material with acceptable cutting speeds, reasonable tool life, and in a cost-effective manner for high-volume production calls for a tool design that has:

- multi-cutting edges;
- an accurate tool shank and interface;
- precision-ground cutting edges;
- separate coolant passages with outlets directed to every cutting edge; and
- an easily replaceable and user-friendly tool head.

The workpiece shown in Figure 11-16 is the common rail, part of a new innovative fuel-injection system for advanced diesel engines. Only a tooling system featuring the aforementioned characteristics can achieve the machining results shown.

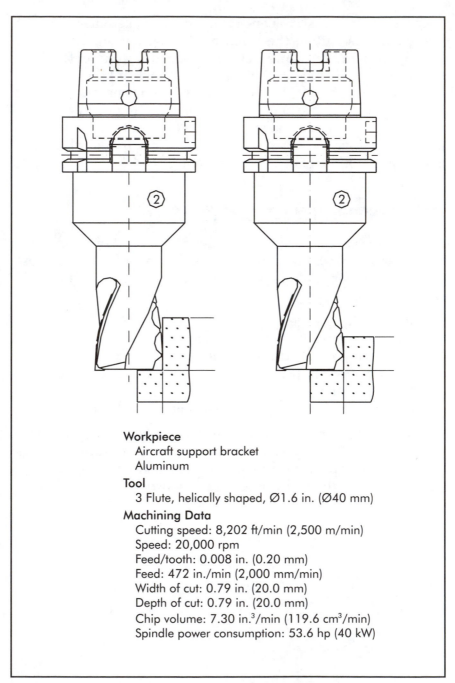

**Workpiece**
  Aircraft support bracket
  Aluminum

**Tool**
  3 Flute, helically shaped, Ø1.6 in. (Ø40 mm)

**Machining Data**
  Cutting speed: 8,202 ft/min (2,500 m/min)
  Speed: 20,000 rpm
  Feed/tooth: 0.008 in. (0.20 mm)
  Feed: 472 in./min (2,000 mm/min)
  Width of cut: 0.79 in. (20.0 mm)
  Depth of cut: 0.79 in. (20.0 mm)
  Chip volume: 7.30 in.$^3$/min (119.6 cm$^3$/min)
  Spindle power consumption: 53.6 hp (40 kW)

*Figure 11-15. Machining aircraft support brackets.*

Workpiece: Common rail
Material: Hardened steel, 62 HRC
Tool: Head-fitting system with 4 cubic-boron nitride (CBN) blades
Machining data: $V_c$ = 312 ft/min (95 m/min),
                    $V_f$ = 8.86 in./min (225.0 mm/min)
Roughness: $R_z$ = 0.04–0.05 mil (1.0–1.3 μm)
Tool life: 1,800 bores

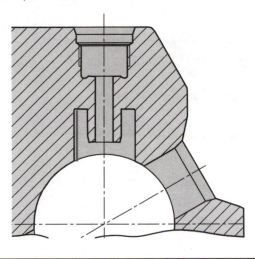

*Figure 11-16. Hard-machining of a common rail.*

# Trends and Outlook 12

Manufacturing processes must be optimized if progress is expected on the manufacturing floor. It is not enough to fix just one component and leave the others untouched. It is not sufficient to acquire the most technologically advanced machine and equip it with average tooling. Nor is it to provide advanced machine and tooling and overlook the tool and part clamping. It is simply a waste of effort if advanced equipment is not run with the optimum machining parameters. Processes have to be productive, efficient, and effective. Every component that is part of the manufacturing process equation has to be at its best. On the path to high-performance machining, there are certain milestones along the way. They constitute trends and advancement.

## MATERIALS

The use of nonferrous metals and composite materials for parts will dominate industry. However, given the demand for extremely hard material, there will still be a need for hardened steels. For example, customers will want materials higher than 50 hardness Rockwell C (HRC). Near-net-shape technology and rapid prototyping will further reduce the amount of tools and machining passes and contribute to high-performance machining through cost, material, and time savings.

## MACHINE TECHNOLOGY

Conceptually, agility will continue to be the driver for judging manufacturing success. Smaller, lighter, and faster machines will be predominant. They will feature single or multi-spindles. They will be capable of doing practically all operations including drilling, reaming, boring, tapping, turning, and even grinding. Part fixturing has to be such that even complex workpieces can be machined with one clamping regardless of the type of operation.

Regular production machines offer spindle speeds up to 20,000 rpm and axes acceleration rates of up to 2.5 g. They can accommodate wet, dry, or semi-dry machining and dispose of chips automatically. Automated part loading and unloading is essential to reduce cut-to-cut time to within split seconds. A built-in, automatic adjustment for imbalances of the entire machining process is provided. So is the use of lasers for alignments and part qualification.

## TOOLING TECHNOLOGY

The future will bring intelligent tooling that can adjust itself for stock variations, surface-finish requirements, tool wear, and application-specific tool coatings. This will be especially true for high-precision machining that involves expensive/exotic workpiece material.

High precision and thoroughly engineered cutting tools have to be developed. Whatever the level of complexity has to be, the equipment must be user-friendly and reliable. High-speed cutting tools, self-centering with cylindrical shafts, can now be clamped easily with high precision through shrink fit. The hollow-shank (HSK) taper interface between machine spindle and cutting tool ensures high repeatability, accuracy, and secures accurate torque transmission. In general, tooling will be at the forefront of technological advancement for ease of handling, standardization, tool life, reliability, and cost-effectiveness.

## PROCESSES

High-performance machining incorporates self-contained processes with clearly defined objectives. The common denominators for high-speed machining are high productivity, high precision, and cost savings.

High-speed/high-velocity cutting has already permeated the metalworking industry. The trend toward dry or semi-dry-machining will be much more prevalent in the future. The technology is available and now is the time to adopt it. Finish machining in one pass is clearly the focus of advanced cutting-tool technology, in conjunction with the workpiece prepared for it. Near-net-shaped workpieces and precisely engineered tooling can forego the need for costly pre- and post-machining processes.

First part, good part, and zero-defect machining require the cutting tool and workpiece to be fine-tuned to the machining process while considering quality, design, technical capabilities, and limits. Because of their interdependence from the view of the process, cutting tool and workpiece have to be considered holistically.

Optimizing high-performance machining means going beyond the traditional method and scrutinizing every angle of the process. The best approach is concurrent engineering, which uses the strengths of teams, techniques, and technology. By using cross-functional teams, manufacturers and suppliers use best practices to determine best processes.

In a global manufacturing world where manufacturers are networked at several locations and with their supply chain, products have to be made uniformly with the same level of reliability and quality. The goal is standardization of processes with some local allowable deviation. Cost, quality, and time have to be maximized. The principles of concurrent engineering have to be expanded upon through concurrent manufacturing, which in essence, stresses process instead of product. Virtual teams, interlinked through internet/intranet, exchange experiences, set uniform standards, and improve upon existing scenarios. These efforts must be ongoing, just as high-performance machining and manufacturing are ongoing.

# Bibliography

*Application of Diamond Films and Related Materials*. 1995. National Institute of Standards and Technology (NIST) Special Publication #885. Washington, DC: U.S. Government Printing Office.

Bakerjian, Ramon and Phil Mitchell, eds. 1993. *Tool and Manufacturing Engineers Handbook*, Fourth Edition. Volume 7: *Continuous Improvement*. Dearborn, MI: Society of Manufacturing Engineers.

———. 1992. *Tool and Manufacturing Engineers Handbook*, Fourth Edition. Volume 6: *Design for Manufacturability*. Dearborn, MI: Society of Manufacturing Engineers.

Chopra, Sanil and Peter Meindl. 2000, December. *Supply Chain Management: Strategy, Planning, and Operations*. Englewood Cliffs, NJ: Prentice Hall.

Dornfeld, David A. 2001. *Research Reports 2000-2001*. Berkeley, CA: University of California, Laboratory for Manufacturing Automation.

Engdahl, Chris. 1998. *CVD Diamond-coated Rotating Tools for Machining Advanced Materials*. Dearborn, MI: National Center for Manufacturing Sciences (NCMS) Workshop Series.

Erdel, Bert P. 1999, Fourth Quarter. "Environmental Issues in Machining." *Machining* technical quarterly. Dearborn, MI: Society of Manufacturing Engineers.

Eversheim, F., T. Pfeifer Klocke, and M. Weck. 1999. *Wettbewerbsfaktor Produktionstechik*. Aachen, Germany: Shaker Verlag GmbH.

Hrone, Steven M. 1993. *Vital Signs: Using Quality, Time, and Cost Performance Measurements to Chart Your Company's Future*. New York: Arthur Anderson & Co.

Hwang, Edward I. 2000. *Research Reports 2000-2001*. Berkeley, CA: University of California, Laboratory for Manufacturing Automation.

Juran, J.M., and Frank M. Gryna, Jr. 1989. *Quality Planning and Analysis*. New York: McGraw-Hill Publishing Company.

Krass, S., A. Gill, and P. Sarid. 2001. *CNC Simplified*. New York: Industrial Press.

Lehmann, Donald R. 2002. *Product Management*, 3rd Edition. New York: McGraw Hill.

Lutz, Bob. 1998. *Guts: The Seven Laws of Business that Made Chrysler the Hottest Car Company*. New York: John Wiley and Sons.

Oleson, John D. 1998. *Pathways to Agility: Mass Customization in Action*. New York: John Wiley & Sons, Inc.

*Research Reports 1998-1999*. 1999. Berkeley, CA: University of California, Laboratory for Manufacturing Automation.

Schrader, George F. and Ahmad K. Elshennawy. 2000. *Manufacturing Processes and Materials*. Dearborn, MI: Society of Manufacturing Engineers.

Schulz, Herbert. 1996. *High-speed Machining*. Munich, Vienna: Carl Haurer Publication Company.

Smith, Graham T. *Advanced Machining: The Handbook of Cutting Technology*. New York: IFS Publications/Springer-Verlag.

VDI. *Cutting Material and Tools*. 1989. Dusseldorf, Germany: VDI-Verlog GmbH.

Weck, Manfred, Walter Eversheim, Wilfried Koenig, and Tilo Pfeifer. 1991. *Production Engineering: The Competitive Edge*. Burlington, MA: Butterworth Heinemann Ltd.

WZL-Aachen. 1999, June. *Production Technology*. Aachen, Germany: Fraunhofer Institute, Shaker Publication Company.

Zimmerman, Fred and Dave Beal. 2002. *Manufacturing Works*. Dallas, TX: Dearborn Trade Publishing.

# Index